# We Are Not Alone

"A compelling landmark book that will take you on an exhilarating ride."

*Daily Express*

"Packed with facts and written in an absorbing style, this book leaves you in no doubt that extraterrestrial microbes could yet be found in our Solar System."

*BBC Sky at Night*

"An exhilarating ride. The arguments advanced for the prevalence of life in the Solar System are well developed and highly stimulating."

*Times Higher Educational Supplement*

"A highly recommended case that brings the new possibility to readers of life at various levels. Conditions which are probably too hostile to permit life to exist - certainly life as we know it...Compelling. The search for extraterrestrial life has entered the scientific mainstream. Balanced and thoughtful."

**Neil F. Comins** – Professor of Physics at the University of Maine and author of *The Hazards of Space Travel*

"Well written, concise, and elegant. I was barely able to put it down."

**Felisa Wolfe-Simon** – Research Fellow, Department of Earth and Planetary Sciences, Harvard University

"Bang up to date and utterly riveting. An absorbing account of one of the greatest unresolved debates in the search for life."

**Lewis Dartnell** – author of
*Life in the Universe: A Beginner's Guide*

"Interesting, accessible, and refreshingly upbeat."

**Seth Shostak** – Senior Astronomer at the SETI Institute and author of *Confessions of an Alien Hunter: A Scientist's Search for Extraterrestrial Intelligence*

# We Are Not Alone

## Why We have Already Found
## Extraterrestrial Life

### Dirk Schulze-Makuch and
### David Darling

ONEWORLD

OXFORD

A Oneworld Book

First published by Oneworld Publications 2010
Reprinted 2010
This edition published by Oneworld Publications 2011

ISBN 978–1–85168–788–6

Typeset by Jayvee, Trivandrum, India
Cover design by DogEared Design
Printed and bound by
TJ International Ltd, Padstow, Cornwall

Oneworld Publications
185 Banbury Road, Oxford, OX2 7AR, England

# Contents

# CONTENTS

# Preface

Since the time of ancient Greece and perhaps well before that, mankind has dreamed of alien worlds and the beings who might inhabit them. But this long period of speculation may be about to end.

In this book we shall argue that there's powerful evidence to suggest we're not alone in the universe and, in fact, that we have company close at hand, within the Solar System. Data collected over the past three decades indicate that one of our neighbouring worlds almost certainly harbours life and several others may do so. That is the central, controversial claim of this book. We may be on the brink of finally proving that extraterrestrial biology exists – right here, in our cosmic midst.

The authors come from quite different backgrounds but share a common belief: that the signatures of life beyond Earth have already been detected. Dirk Schulze-Makuch, of Washington State University, has been at the forefront in recent years of scientific debate about life on Mars, Venus, and Titan. His explanations of spacecraft data in terms of extraterrestrial biology have attracted worldwide media attention. He is also involved with planning for future space missions. Author and astronomer David Darling has written extensively about the new science of astrobiology.

# Acknowledgements

The authors are grateful to the following colleagues for their helpful conversations, correspondence, suggestions, and source material: Sam Abbas, Dale Andersen, Vic Baker, Penny Boston, Athena Coustenis, Alfonso Davila, Detlef Decker, Ben Diaz, James Dohm, Thomas Eisner, Alberto Fairén, Chaojun Fan, Wolfgang Fink, Roberto Furfaro, David Grinspoon, Huade Guan, Ed Guinan, Victor Gusev, Shirin Haque, Joop Houtkooper, Louis N. Irwin, Mohammed Islam, Gil Levin, Darlene Lim, Jere Lipps, Giles Marion, Chris McKay, David McKay, Anthony Muller, Dorothy Oehler, Marina Resendes de Sousa António, Ed Sittler, Bob Shapiro, Carol Turse, Corby Waste, and Jacek Wierzchos.

We also thank our editor Mike Harpley for his numerous constructive criticisms and tireless efforts to move the book forward.

Finally, and mostly importantly, we are grateful to our families for their support and patience over the two years this book went from first thoughts to finished manuscript.

# Illustrations

# Chronology of the Quest for Alien Life

165    Lucian of Samosata writes his *True History*, in which extraterrestrial life is talked about (whimsically) in detail for the first time.

*c.*1450  Nicholas of Cusa revives the ancient idea that planets might be inhabited.

1650   Christiaan Huygens observes Syrtis Major, the first permanent feature observed on another planet (Mars).

1859   Charles Darwin publishes *The Origin of Species*.

1862   French astronomer Emmanuel Liais suggests that dark areas on Mars might be vegetation.

1877   Italian astronomer Giovanni Schiaparelli reports seeing what he calls "canali" (channels) on Mars.

1893   British biologist James Reynolds suggests that life could be based on silicon.

1894   Percival Lowell founds his observatory in Arizona and publishes his first book, *Mars*.

1897   H. G. Wells publishes *War of the Worlds*.

1918    Swedish chemist Svante Arrhenius publishes his ideas about life on Venus.

1929    British geneticist J. B. S. Haldane speculates about the chemical origin of life.

1932    Carbon dioxide found in the atmosphere of Venus.

1937    Dutch-American astronomer Peter van der Kamp begins the search for extrasolar planets.

1943    Dutch-born American astronomer Gerard Kuiper discovers the atmosphere on Titan.

1953    Stanley Miller and Harold Urey carry out their famous origin-of-life experiment.

1954    J. B. S. Haldane suggests the possibility of ammonia-based life.

1958    Panel on Extraterrestrial Life, formed to discuss the problem of what extraterrestrial life might be like and how to look for it.

1959    First grant awarded by NASA, to Wolf Vishniac, to develop a prototype alien life detector.

1960    NASA begins planning to send a spacecraft to search for life on Mars.

1962    Mariner 2 flies past Venus.

1965    Mariner 4 flies past Mars.

1967    Soviet probe Venera 4 enters the atmosphere of Venus.

1970    Amino acids found in the Murchison meteorite.

1971    Mariner 9 goes into orbit around Mars.
        American astronomer John Lewis suggests the possibility of an underground ocean on Europa.

1976    Viking 1 and 2 land on Mars.

1977    Hydrothermal vent communities found on Earth's ocean floor.

1985    Experiments show that bacteria could survive in space.

1988    First tentative detection of an extrasolar planet which was subsequently confirmed (in 2003).

1995    First definitive detection of an extrasolar planet around a Sun-like star, 51 Pegasi, by Swiss astronomers Michel Mayor and Didier Queloz.

1996    Claim of Martian "fossils" and other biogenic remains in Martian meteorite ALH 84001 by David McKay and colleagues.
        Data from the Galileo probe suggest the presence of an underground ocean on Jupiter's moon Europa.

1999    First multiple planetary system found around a Sun-like star (Upsilon Andromedae).

2000    Mars Global Surveyor finds ancient water channels on Mars.

2004    The Mars Exploration Rovers *Spirit* and *Opportunity* land on Mars to explore its surface geology.
        Discovery of methane on Mars.

2005    Images from the Mars Global Surveyor show the formation of a new gully between 2001 and 2005, indicating the presence of liquid water on Mars still today.
        Huygens probe lands on Titan.

2006    The Cassini orbiter discovers lakes of methane and a hydrological cycle of methane on Saturn's moon Titan.

2007    First Earth-like extrasolar planets (Gliese 851c and 851d) found within the habitable zone of their star.

2008    Phoenix Lander touches down on Mars.

2009    Kepler launched on its mission to identify Earth-like extrasolar planets.

2011    Mars Science Laboratory scheduled for launch.

2018    ExoMars scheduled for launch.

# Introduction

I s Earth unique? Is life somehow special to this planet, or is it widespread throughout the cosmos? And, if there is life else-where, what is it like? Philosophers and poets alike have grappled with these questions for centuries. Today we are still intrigued by the possibility of extraterrestrial life. The difference now is that we have real data to work with. We've become familiar with some of our planetary neighbours, seen them close up, even in some cases landed on them and sampled their soil.

The quest for life beyond Earth has shifted from the realm of pure speculation into that of mainstream science. More than 2,000 years ago, Greek thinkers began the debate about the "plurality of worlds" – whether other Earths exist out there. In the first century A.D., the Syrian satirist Lucian of Samosata wrote an entire book about weird and wonderful life forms on the Moon, the Sun, and elsewhere. With the invention of the telescope, speculation intensi-fied. Kepler and Galileo both thought the Moon might be inhab-ited. But, as time went on, scientific attention focused more and more on the planet Mars as the likeliest abode of life beyond Earth. Changes on its surface were glimpsed through eyepieces, suggestive of the seasonal comings and goings of vegetation. Some observers thought they could see evidence of water channels on the surface of the Red Planet – a claim that helped fire public interest in nearby alien life and even the prospect of other-worldly intelligence.

To some extent there has always been a disconnection between popular and scientific expectations about life elsewhere. By and large over the past century, public opinion has been enthusiastically positive, while the hopes of scientists have waxed and waned almost as seasonally as the colour changes on Mars. At the dawn of the space age, hopes were high of finding at least microbial life on Mars and possibly even lichen or more advanced vegetation. In this period of optimism, mankind's first (and so far only) *in situ* life-detection mission – Viking – was conceived. But by the time it arrived at the Red Planet, biological expectations had plummeted because of the bleak, Moon-like pictures returned by the pioneering Mars flyby probe, Mariner 4. It was against the backdrop of this negative mind-set that the results of the Viking biology experiments, which initially seemed to point to the presence of life, were interpreted. A consensus emerged that Viking had drawn a blank. But had it? In the decades that followed, the Viking results have been reinterpreted, and there have been recent suggestions that they may indeed point to life, though of a very different order to that with which we're familiar.

Just as the debate about Martian life has been thrown wide open, so has the case for astrobiology elsewhere in the Solar System. Unexpected potential locales for life have turned up on icy moons orbiting Jupiter and Saturn and even in the atmosphere of Venus. The evidence is stacking up that, even within the Sun's domain, we have company. Meanwhile, the greater universe beckons, with the prospect of billions of uncharted worlds and life of a diversity we can barely begin to grasp.

We invite you now to consider why we are not alone.

# Part I

There's no place like Mars – for the astrobiologist, that is. It isn't that Mars stands alone, among the worlds we know, in its prospects for supporting life. Mars is special, from the alien-hunter's point of view, due to a combination of factors. It's close at hand (the second-nearest planet to Earth, after Venus). It has an aura of mystery about it because viewed through telescopes its surface seems to change subtly from one season to the next. Environmentally, it's the most Earth-like place we know, and, spacecraft observations have revealed, it was even more Earth-like in the past. Not least, it's the subject of numerous science fiction stories and speculative science works going back well over a century.

It would be surprising and disappointing if incontrovertible proof of past or present life were not found on the fourth planet from the Sun. It would seriously downgrade our hopes that life is reasonably common throughout space and time, because this is a world where, by every indication, organisms of some kind *should* exist. Fortunately, the evidence for Martian life is excellent. We are not alone, we shall argue, even within a cosmic stone's throw of our home world.

# 1

# Life Signs

There was a time when talk of life beyond Earth would have branded you a philosophical daydreamer, a writer of fiction, or an eccentric, but certainly not a true scientist. After all, where were the data? Science thrives on what it can observe. But who had ever seen a Mercurian albatross, a Venusian orchid, or a giant Jovian cloud-dweller? Where was the barest inkling of alien biology?

True, there'd been no shortage of fanciful speculation about life "out there." For more than two millennia, folk had wondered whether the Moon and other worlds might have their own home-grown flora and fauna. The nineteenth century saw a philosophical and theological debate rumble about how often, if ever, life took hold elsewhere in the cosmos. Following the birth of Darwinism, and the growing acceptance among biologists that life had evolved naturally here on Earth, that debate intensified and bled into the margins of science. After all, if organisms had sprung up here and flowered into a multitude of forms, why not also on Venus, Mars, and a million other worlds throughout space? Why shouldn't Darwinian evolution be universal in scope?

## Rise of the Martians

By the last quarter of the nineteenth century, astronomers and fiction writers alike had honed in on the Red Planet, Mars, as the likeliest abode of life in our cosmic neighbourhood. It was smaller, cooler, and drier than Earth, and its atmosphere was a good deal thinner. Yet it was by far the least hostile-seeming place by fragile human standards. Some who had eyed the planet keenly through their telescopes claimed it had features that looked suspiciously like water-courses. The respected Italian astronomer Giovanni Schiaparelli was among those who thought he saw these markings. He called them *canali*, which means nothing more presumptuous than "channels." But inevitably the word was mistranslated as the much more loaded term "canals." That hint of artificiality, though never intended, was all the encouragement that American business-man-turned-astronomer Percival Lowell needed to fuel an obsession that would last the rest of his life.

Almost single-handedly, as the 1800s drew to a close, Lowell fanned a wildfire of public fascination with the possibility of advanced Martians. He invented an alluring saga of the canals of Mars and an ancient race which, beleaguered by drought, had built artificial channels to bring meltwater from the polar caps to irrigate the dry desert regions of their dying world. During hundreds of hours spent at the eyepiece, Lowell mapped scores of lines criss-crossing the Martian surface with geometric precision. He saw, from his purpose-built observatory in Flagstaff, Arizona, oases where the linear markings intersected. And he wrote three marvellous, imagination-capturing books (published between 1895 and 1910) about Mars and the creatures that dwelled there, which were eagerly consumed by an enthralled popular audience.

Even as Lowell's startling claims found resonance with a public gripped by Mars fever, in England H. G. Wells published his dark vision of Martian invaders in *The War of the Worlds*, first serialised in 1897. The notion of an alien race, technologically ahead of us and determined to usurp our planet, proved irresistible to the layperson. As late as 1938, Orson Welles was able (unintentionally) to trigger mass panic in the United States with a masterful radio dramatisation of his near namesake's novel. Meanwhile, Edgar Rice Burroughs made a small fortune through sales of his lightweight Mars romances – ten of them in all, starting with *The Princess of Mars*, published between 1917 and 1948. Even into the fifties and beyond, film producers were able to milk the intelligent-life-on-Mars theme with a series of B movies that ranged from the surprisingly good to the truly awful.

The fact is, the vast majority of people *wanted* to believe in life on Mars, in life throughout space, and, preferably, in brainy extraterrestrial life, even if it wasn't always friendly, because that belief made the universe seem a much more interesting place. But science is more cautious. It demands solid bedrock on which to build its theories. With life only known to exist on Earth, what basis was there for supposing that other planets might be inhabited?

## Birth of a science

Unlike physics and chemistry, which have long been considered universal in scope, biology has traditionally been a more parochial, one-planet science. For most of its long history, biology has been synonymous with terrestrial biology. Any professional biologist who openly speculated about other kinds of life tended to be given short shrift by his or her colleagues. Yet, in recent times, a new and respectable branch of the science has opened up, called astrobiology, which aims to extend biological theories and knowledge to the universe at large. As well as seeking out life on other worlds, astrobiologists hope to arrive at a broader definition or concept of life that covers all the forms it might take and the various ways it might evolve and originate.

With this last point in mind, we can trace the roots of astrobiology back to the man who penned *The Origin of Species*. In 1871, in an uncannily prescient letter, Darwin wrote:

> It is often said that all the conditions for the first production of a living organism are now present, which could ever have been present. But if (and oh! what a big if!) we would conceive in some warm little pond, with all sorts of ammonia and phosphoric salts, light, heat, electricity, etc., present, that a protein compound was chemically formed ready to undergo still more complex changes. At the present day such matter would be instantly devoured or absorbed, which would not have been the case before living creatures were formed.

Few of Darwin's scientific peers went along with this "warm little pond" idea – that life had somehow emerged from a chemical soup. It was too fantastic and far ahead of its time. But the idea was revived in the 1920s independently by two biochemists, J. B. S. Haldane in Britain and Aleksandr Oparin in the Soviet Union. While Haldane's and Oparin's theories differed in detail, their gist was the same: Earth's dawn atmosphere was totally unlike the one we breathe today. It was rich in reducing gases such as ammonia and hydrogen, but had little or no free oxygen. When reducing gases come into contact with substances containing oxygen they tend to capture the oxygen; for example, in a reducing atmosphere carbon dioxide would have its oxygen stripped away, leaving the carbon available to react in new ways. A reducing, oxygen-free atmosphere, reasoned Haldane and Oparin, would give organic (carbon-bearing) molecules the chance to build up in the sea and grow more complex until, after millions of years, these macromolecules became organised into the first primitive cells.

Other biochemists at the time, doing seminal lab work to retrace the earliest steps leading to life, persisted with oxygen-rich environments. It wasn't until the 1950s that origin-of-life experimenters started to take Darwin's "warm little pond" and the reducing atmosphere scenarios seriously. The shift came because scientists realised that a key to getting life off the ground was the creation of amino acids, the building blocks of proteins. They began to suspect that amino acids might assemble spontaneously in water that was in contact with a reducing atmosphere. That belief was triumphantly vindicated by an experiment done in 1953 by Stanley Miller, a doctoral student of the Nobel Prize-winning chemist Harold Urey. Miller zapped a flask containing methane, ammonia, water vapour, and hydrogen – Oparin's recipe for Earth's primitive atmosphere – with sparks between tungsten electrodes. The fallout from the mini electrical storm was allowed to collect in a smaller flask of water. After a week, a murky brown scum had gathered on the surface of this sea-in-a-bottle. And in the mysterious goo, Miller found, were all sorts of intriguing organics, including tiny amounts of the amino acids glycine and alanine, crucial ingredients of all terrestrial life.

Six years later, emboldened by their lab results, Miller and Urey expanded their remit to the cosmos at large. They started talking openly about life beyond Earth. "The planets Jupiter, Saturn, Uranus and Neptune are known to have atmospheres of methane and ammonia," they pointed out. Also, said Miller: "The atmosphere of Mars would have been reducing when this planet was first formed, and the same organic compounds would have been synthesised in its atmosphere. Provided there were sufficient time and appropriate conditions of temperature, it seems likely that life arose on this planet."

A generation earlier, Miller's comments on Mars might have gone unheeded. But now his hypothesis was scientifically relevant, because there would soon be the means to test it. The space age had begun. On 4 October 1957, Sputnik had rocketed into orbit and transformed humankind's dreams of flying to other-worldly bodies into hard-edged reality. Before long, spacecraft would be on their way to Mars to see it up close and then to stand on its surface. If, as Miller suggested, life had arisen there, and if it had survived to this day, it ought to be possible to sniff it out.

But there were new dangers to be faced as interplanetary probes prepared to set sail. What if the spacecraft sent to look for life on other worlds carried with them unwelcome visitors? We're painfully aware of how vulnerable we are to new strains of bugs, to which our bodies have little or no resistance. What if our probes transported bacteria which contaminated other worlds and decimated any native populations? At the very least, terrestrial hitchhikers might wreck any efforts at extraterrestrial life detection.

Geneticist and Nobel Prize-winner Joshua Lederberg was among the first to grasp, in the wake of Sputnik, that planetary protection had to be a top priority when exploring other worlds of the Solar System. His concerns soon won the support of the National Academy of Sciences and then of the newly-formed National Aeronautics and Space Administration (NASA). By late 1959, NASA made it clear that it planned to sterilise, as much as technology would allow, any spacecraft that were going to fly close by or land on the Moon or planets. At the same time the agency had started looking into how it should engage with the life sciences. Two areas stood out as being of prime importance: space medicine and extraterrestrial biology.

With manned spaceflight imminent, it was crucial to understand how humans would react to the stresses of launch and re-entry, and to the untested realm of weightlessness. Using giant centrifuges, pressure chambers, rocket sleds, and the like, space medicine focused on the physiological and psychological effects of space travel by exposing test subjects to conditions similar to those they might encounter on a real journey into orbit. Future plans also called for the robotic exploration of potentially life-bearing worlds, including Mars and Venus. For this reason, priority was also given to understanding what kinds of life might be encountered and how to mitigate any problems that might arise from terrestrial organisms coming into contact with their alien counterparts.

In August 1960, the hunt for extraterrestrials began in earnest. NASA authorised the Jet Propulsion Laboratory (JPL) in Pasadena, California, to study an extraordinary mission: to land a capsule on Mars and begin the search for life there.

## On the trail of ET

But how exactly do you hunt for aliens? Spotting anything large, of course, would be child's play. Any Martian spiders or even lichen-encrusted rock would be easy to see on the surface with a TV camera. In fact, scientists in the fifties and early sixties still considered vegetation on Mars a distinct possibility, given the curious seasonal waves of darkening (now known to be caused by dust storms), visible through telescopes, that spread out from the polar caps as the ice retreated each spring. Yet, realistically, the best chance for life seemed to be in the form of microbes in the soil. So, the question was: how do you go about trying to detect microscopic bugs, of an unknown nature, with an unmanned probe tens of millions of kilometres away?

In March 1959, NASA handed out its first grant, for the princely sum of $4,485, to answer that question. It went to Wolf Vishniac, a Latvian-born microbiologist at Yale, to develop "a prototype instrument for the remote detection of microorganisms on other planets."

Vishniac was under no illusion about the scale of the task he faced. He and 18 other scientists, forming the Panel on Extraterrestrial Life, had met in Cambridge, Massachusetts, in December 1958, to thrash out the problem of what kinds of life, if any, they might reasonably expect to find away from their own planet.

It might be, realised the Panel, that life was unique to Earth. Or perhaps life had once flourished on other worlds, only to succumb to catastrophic changes in the environment to which it couldn't adapt. But if there were creatures alive today on other planets in the Solar System, what might they be like?

Vishniac and his colleagues reckoned there were four main possibilities. The first was that living things elsewhere were pretty much the same as those found on Earth – same chemistry, same overall appearance. That could be because they had a common point of origin and there'd been cross-fertilisation between worlds (for example, via meteorites). Or maybe life just always evolved the same way. The second possibility was that alien life had the same underlying chemistry but was different in other ways because different environments, past and present, had affected the course of its evolution. Scenario three was that extraterrestrial life was different even at the chemical level. It might be based, for example, on silicon rather than carbon. Finally, life on another world might be very primitive, no more evolved than the earliest ancestors of today's terrestrial microbes. Few of those pioneering astrobiologists, gathered in Cambridge at the end of 1958, fervently backed any one of these possibilities over any other.

How does a scientist detect something if he doesn't know what form it might take? Vishniac and his colleagues made some fundamental assumptions, one of which was that life elsewhere would have a carbon base. "It may turn out that we are deluding ourselves," said Vishniac, "that we are simply limited in our imagination because of our limited experience." But the fact remained that, as far as biochemists were aware, nothing came close to carbon in its chemical versatility and its talent for making large, stable molecules.

Assuming that all biochemistry is carbon-based still leaves the would-be alien hunter with a big unanswered question: what is life? What defines it? Again the problem is we have only one instance of life – Earth life – to go on. That situation hasn't changed since the dawn of astrobiology, or "exobiology" as Lederberg first called it, in the late fifties. What has changed is that astrobiology has become a practical science, involving real instruments and probes. This shift to a hands-on approach began with the early work at NASA to design and build alien life detection gear for the mission to Mars that became known as Viking.

## Troubled times

Detailed planning for the surface exploration of Mars took place in the 1960s in the midst of major political tensions and a sudden, rapid expansion of human activities in space. The sixties saw the Vietnam War reach its most deadly stage and the Cold War threaten us with global annihilation. Meanwhile, the Apollo programme, a direct response to the challenge of Soviet Communism, was in full swing, heading toward its spectacular goal of putting humans on the Moon by the end of the decade. But, on a less optimistic note, the prospects for life on Mars seemed to fade dramatically in the wake of photos sent back by the Mariner 4 flyby probe in July 1965. Those grainy, black-and-white images revealed a stark, Moon-like landscape pockmarked by craters and with little sign of activity or of the watery channels that some astronomers thought they'd glimpsed through telescopes. Later it turned out that Mariner 4 had been unlucky in that its cameras just happened to be trained on some of the most desolate regions of the Red Planet. But, at the time, its findings raised serious doubts about whether it made sense to pursue the quest for living Martians.

Arguments against a search for life on Mars were made not only on scientific grounds, but also from political, social, economic, and even religious standpoints. Philip Abelson, a physicist and editor of the journal *Science*, was among critics who dismissed the economic case for exploring the Moon or Mars and pointed to a need to tackle instead more pressing problems here on Earth. On the other side of the fence, those supporting the search for extraterrestrial life insisted it was the most exciting, challenging, and profound issue to emerge from Western thought in 300 years.[1]

Despite the nay-sayers and the seemingly gloomy news from Mariner 4, the Voyager Mars Program was born. It was to have been part of the Apollo Applications Program (AAP), intended to use surplus hardware from the manned lunar landings. Twin orbiters – very much like the Mariner 9 probe (see figure 1) which sent back more than 7,000 images from Martian orbit between 1971 and 1972 – and landers patterned on the successful lunar Surveyor craft would be launched together, in 1974–75, by a giant Saturn V rocket. They would help pave the way for a human expedition to Mars in the 1980s. But the project never got off the drawing board.

As Congressional impatience with the soaring price-tag of Apollo grew, funds that had been set aside for Voyager Mars, and the AAP as a whole, were diverted to keep the Moon program on track. In 1967, Voyager Mars was axed altogether. But shortly after, a new, cheaper scheme to achieve essentially the same objectives was hatched at NASA's Langley Research Center. Like Voyager Mars it involved twin orbiters and landers. But instead of these craft being launched, somewhat riskily and expensively together, by a single Saturn V, they would be sent on their way as single orbiter-lander combinations by smaller Titan rockets.

**Figure 1** Mariner 9 was the first spacecraft to orbit another planet and transmitted thousands of images from Mars

This reworked concept was formally proposed and given a new project name: Viking. Jim Martin from Langley became Project Manager (later Mission Director) and Jerry Soffen was chosen as Viking Project Scientist.

Soffen had earned his PhD at Princeton University and worked for a while at the New York University School of Medicine before joining NASA-Caltech's Jet Propulsion Laboratory in Pasadena, California. There he'd become one of the first professional astrobiologists, devising instruments for the detection of life on Mars. He'd served as deputy project scientist on the Mars Voyager program before its cancellation. Now, he, Jim Martin, and Thomas Young, the science integration manager, were faced with a tricky task: to pick a science team for Viking.

What they wanted were team players – people who could work closely and cooperatively together. They also needed researchers with a broad scientific background who were ready to commit to a project that might run for many years. Of the coterie of biologists with the necessary skills, not everyone was interested in extraterrestrial biology or prepared to risk their career by getting involved with it. However, by February 1969, a preliminary list of Viking science team members was drawn up, which was later finalised. It quickly became clear, however, that the chosen researchers weren't going to gel into the harmonious group that managers had hoped for. Bickering between team members was to become a familiar backdrop against which the Viking drama unfolded.

## A lab out of this world

A set of life-detection experiments to be carried to Mars by the Viking landers also had to be chosen from around 150 proposals submitted to NASA. Eventually, four pieces of equipment were singled out. These were the Gas Exchange (GEX) experiment, the Labeled Release (LR) experiment, the Pyrolytic Release (PR) experiment, and the Wolf Trap. The last was named after Wolf Vishniac and based on the detector he'd developed with the first-ever NASA grant for astrobiology, issued back in 1959. All of the experiments had to be shoehorned into a container less than 27 litres in volume (equivalent to a square box measuring just 30 centimetres, or about one foot, on each side) and weighing less than 16 kilograms (35 pounds). This miniature biology lab would be supplemented by an instrument known as the Gas Chromatograph Mass Spectrometer (GCMS) whose main task was to track down and identify organic molecules in the Martian soil.

As its name suggests, the GCMS was a combination of two devices – a gas chromatograph and a mass spectrometer. The first of these could separate out from a sample of soil any substances that could be turned into a vapour by heating. The second could determine the molecular mass of each of the vaporised substances and hence its likely identity. Together, the gas chromatograph and the mass spectrometer were designed to identify many compounds with great certainty and at very low concentrations. On the other hand, the GCMS did have some limitations. For example, organic compounds that didn't turn into a gas when heated to a set temperature, or that reacted with other substances upon heating, would go undetected.

The four experiments chosen for Viking differed in what they assumed about the surface environment on Mars, especially the availability of liquid water, which was (and, to some extent, still is) a major unknown. They also differed in their approach to encouraging any kind of native life to show itself.

The Gas Exchange experiment was headed by Vance Oyama, a biochemist who was chief of NASA's Exobiology Branch at the agency's Ames Research Lab. Before Viking blasted off, Oyama was one of the scientists most upbeat about the chances of finding life on Mars. He thought that Mars, like Earth, might be blanketed with microbes, including plenty of heterotrophs – organisms that feed on a ready-made supply of organic materials. This concept of a planet-wide community of microbes on Mars similar in abundance to soil bacteria on Earth came to be known as the Oyama model. The Gas Exchange experiment was designed to test the Oyama model by taking a sample of Martian soil, exposing it for 12 days to water or water vapour, nutrients, and a simulated Mars-like atmosphere of mostly carbon dioxide, with some helium and krypton added. It would then check several times during the incubation period for any sign of the uptake or release of gases that might point to metabolic activity of life in the soil. The experiment could be run in either of two ways. The first, known as "humid" mode, assumed that any microbes in the soil needed only a dash of water vapour in the atmosphere to kick-start their metabolism. Incubated with nothing more than a gassy mixture of carbon dioxide and water vapour they would be encouraged to become active.

In "wet" mode, the Gas Exchange experiment assumed that whatever Martian organisms were present needed to be given a nutritious meal as well as some water to spring into action. The alien diners would be tempted with a hearty broth, nicknamed "chicken soup," containing 19 different amino acids (the building blocks of proteins), vitamins, a number of other organic compounds, and a sprinkling of inorganic salts.

A lighter version of chicken soup would be served up by the Labeled Release experiment, devised and led by the industrial engineer Gilbert ("Gil") Levin. Each of the organic substances in the watery Labeled Release concoction – formic acid, glycine, glycolic acid, D-lactic acid, L-lactic acid, D-alanine, and L-alanine (D and L standing for left- and right-handed versions of the same compound) – contained atoms of radioactive carbon-14. The hope was that

Martian microbes would metabolise the "labelled" nutrients and then give off carbon dioxide containing carbon-14, which could be easily detected by a radioactivity counter.

The Pyrolytic Release experiment, headed by Norman Horowitz, a geneticist at the California Institute of Technology, was aimed at finding not heterotrophs, which feed on ready-made organic matter, but phototrophs. These are organisms, like plants and some types of bacteria, which can make their own complex organics from just sunlight, carbon dioxide, and water vapour. The Pyrolytic Release experiment would nurture a soil sample with these three simple ingredients, using carbon dioxide laced with radioactive carbon-14. If photosynthetic organisms were in the soil they'd "fix" some of the carbon-14; in other words, they'd incorporate it into their cells during metabolism. After a few days, whatever gases were left in the incubation chamber would be removed and the soil sample heated to a sterilising temperature of 650°C (1200°F). This fierce heating would drive off any carbon-14 that had been converted to biomass, rendering it detectable by the radioactivity counter.

The fourth experiment, the Wolf trap, was built to take a small sample of soil, inundate it with a relatively large volume of water, and then monitor the sample for any increase in cloudiness (turbidity) that might be due to the action of microbes. Of all the experiments, it called for the most amount of water – a factor not in its favour, given the mounting evidence from Mariner 4 and other pre-Viking probes that Mars was desperately dry.

## Then there were three

Right from the start, Viking was, like many space missions, plagued by financial problems. A lot of its cost overruns had to do with the fiendish complexity and technical difficulties of the biology

17

experiments and the Gas Chromatograph Mass Spectrometer. The life-detection gear alone had 40,000 parts, including tiny ovens to heat the samples, little ampoules holding nutrients which had to be broken unerringly on command, bottled radioactive gases, Geiger counters, some 50 valves, a xenon lamp, and a maze of transistors. At a specially convened crisis meeting Jim Martin told the mission scientists that costs would have to be slashed by another \$17 million. There was a mounting chorus, from outside the life sciences group, to drop the Gas Chromatograph Mass Spectrometer and the biology payload altogether.

Harold ("Chuck") Klein, head of the biology team, protested vehemently. He insisted that the landers should take priority over the orbiters and that it would be better to discard the orbiter imaging system than take away anything from the experiments to make direct measurements on the Martian surface. After all, this was the very first attempt to learn something *in situ* about conditions at ground level on the fourth planet. Klein's argument won the day and the life-seeking instrumentation was saved. But, reflecting unease among NASA management about the potential for catastrophic failure of the surface mission, in July 1971 Jim Martin issued Viking project directive number 6: "It is project policy that no single malfunction shall cause the loss of data return from more than one scientific investigation." This was clearly aimed at the biology payload since each of the biology experiments was considered to be one scientific investigation and there were many possibilities for single point failures.

To help economise, the biology payload was simplified and part of the design cut out altogether. But that wasn't enough for Jim Martin. With a couple of years to go before launch, one experiment, he concluded, had to be ditched entirely to keep down the size, weight, and electrical power requirement of the biology payload.[1]

It fell to Harold Klein, Joshua Lederberg, and Alex Rich, all biology team members who weren't associated with any particular experiment, to make this thankless decision. In the end they chose to sacrifice the Wolf Trap because, of all the experiments, it seemed to make the least Mars-like assumption with regard to water availability. The other experiments were more sparing in their use of water – a crucial fact in their favour, based on the depressing (but, as it turned out, misleading) pictures sent back by Mariner 4. The Pyrolytic Release experiment assumed that the Martian environment was very dry and called for no water to be added at all in its first round of tests. Only a drop of water would be introduced in the Labeled Release experiment, while in the Gas Exchange experiment just enough nutrient solution would be injected to moisten the sample.

Not surprisingly Wolf Vishniac was bitterly disappointed. He complained that Lederberg and Rich weren't familiar with the current status of his experiment. He also claimed, after making some inquiries, that there'd been earlier clandestine talk about dropping his experiment which had prejudiced the final choice. Project management and NASA headquarters both stressed that they wanted Vishniac to stay on as part of the Viking mission group, but that came as little consolation to him. The biology team closed ranks and protested the decision to kill off the Wolf Trap, but it was too late.

Vishniac was now in a tough position. He agreed to continue working alongside the other mission scientists. Yet, despite being the assistant biology team leader, he no longer had any NASA funds to support his research projects and staff. He'd been the first person in history to receive government money for astrobiology, but now the funding was gone and he had trouble finding other support. Many scientists not affiliated with NASA shunned him because he was involved with what they considered a wild goose chase for alien life. The National Institute of Health refused him a grant application and told him unofficially that his request had been given low priority because he was "NASAing" around. The National Science Foundation didn't renew one of his grants, partly because of his links with the space agency. Eventually Vishniac told Jerry Soffen that he'd have to cut back his participation in the Viking project in order to spend more time on academic work.[1]

## Life and death in Antarctica

Despite everything, Vishniac continued to pursue his passion. He was determined to explore the Dry Valleys of Antarctica, a gruelling cold desert environment that's among the most Mars-like of all places on Earth. Antarctica's Dry Valleys are typically 5 to 10 kilometres wide and 15 to 50 kilometres long. Most of them have upper reaches that are barren or occupied only by small alpine glaciers. Their temperatures annually average below −20°C but can easily drop to −50°C and rarely get above the melting point of ice. Water vapour concentrations are extraordinarily low and what limited snowfall there is usually sublimes − in other words, turns directly into vapour without ever becoming liquid.

Vishniac wasn't the only scientist encamped in the Dry Valleys of the southernmost continent in the late 1960s and early 1970s, in the run up to Viking's launch. Norman Horowitz, principal researcher with the Pyrolytic Release experiment, spent five years looking for microorganisms in the region only to draw a blank. In a paper published in 1972, he and his team concluded that it harboured the only truly sterile soil on Earth.[2] According to Horowitz, who'd used the then-standard clinical lab techniques for isolating medical bacteria, there wasn't a snifter of microbial life in all of the Dry Valleys. And on that basis, he was inclined to think that Mars would prove lifeless as well.

Vishniac couldn't have disagreed more. In his view, Horowitz had used nutrient media that were too rich and had therefore killed the very organisms he was trying to detect. In the same year − 1972 (three years before the launch of Viking 1) – that Horowitz went public with his negative results, Vishniac and a graduate student, Stanley Mainzera, used a version of the Wolf Trap to successfully isolate colonies of microbes from the same Antarctic soils that Horowitz had declared lifeless. The trick to Vishniac's success was in serving up a broth in the Wolf Trap that was more to the liking of organisms that lived in a nutrient-starved environment. Vishniac was pretty sure that a rich soup like that in Horowitz's experiment would do the same thing to any Mars bugs that it had done to the microbes in the Antarctic soil – wipe them out with a mixture that was effectively toxic to them.

In retrospect, another question mark about the ability of the Viking instruments to detect life signs emerged from those early studies in the Antarctic. Just as Horowitz failed to find any trace of microbes in the Dry Valley soils, so too, in some cases, did a version of the Gas Chromatograph Mass Spectrometer similar to the one that was bound for Mars. This prototype GCMS found no trace of organic matter in certain Antarctic samples now known to contain living, multiplying microbes. If life existed on Mars but in the low concentrations typical of the Dry Valleys, the GCMS might be blind to it.

Vishniac might have gone on to play a pivotal role in the Viking mission to Mars. Even though his own experiment had been axed, his optimism, scientific acumen, and talent for finding compromises could have made him an invaluable part of the team. His detection of life in the Antarctic Dry Valleys would later be confirmed by other scientists, but one can only speculate on any thoughts and suggestions he might have had about the results sent back by the Viking spacecraft. How would he have interpreted the results of the Viking biology experiments? What would he have thought about the data returned by the Gas Chromatograph Mass Spectrometer? Would he have known how to tweak the experiments to better distinguish between a response caused by living things and one that was due simply to chemicals in the soil? We shall never know. During an expedition to the Dry Valleys in December 1973, to learn how the hardy Antarctic microbes obtain their life-sustaining water and nutrients, Vishniac met a tragic end. Alone, on a steep slope in the Asgard Mountains, he slipped and fell to his death.

## Creatures that shouldn't be

Vishniac suspected that any kind of rich "chicken soup" delivered from Earth might poison microbes adapted to the harsh conditions on Mars. What he didn't know was that there were such creatures living right here on our own planet.

The nutrient-rich tonic of the Gas Exchange experiment was tested, in the run up to the Viking mission, on various samples of Earth soil, from regions as diverse as Antarctica, the Gobi Desert, and Alaska. Most of these samples reacted positively, showing unmistakable signs of microbial metabolism, under both aerobic (oxygen-rich) and anaerobic (oxygen-poor) conditions. But two of the terrestrial soils registered as biologically inert in the Gas Exchange tests. These seemingly sterile samples were called Geyserville (after the Geyserville hot springs region of California) and Antarctica 500. Both came from environments that today are recognised as being among the most Mars-like places on Earth. And both, we now know, are not lifeless at all but are instead inhabited by microscopic fauna of a type wholly unsuspected before Viking blasted off. These strange microbes are *extremophiles* – organisms adapted to life under very different conditions than those which suit humans and most other animals and plants.

Unbeknownst to Gas Exchange experiment leader Vance Oyama, the Geyserville sample harboured *thermoacidophilic* (heat-and-acid-loving) microbes which like nothing better than a strong acid bath and temperatures above 60°C (see figure 2). They most

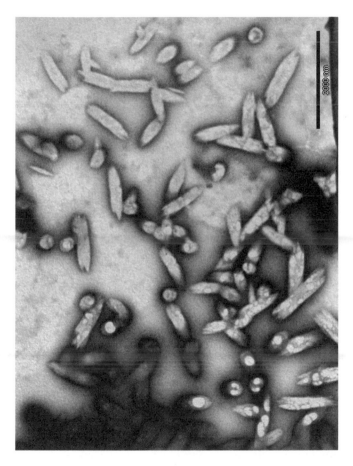

**Figure 2** Heat and acid-loving sulphur bacteria from a hot spring location similar to Geyserville

commonly extract energy from sulphur or iron compounds – a "mineral soup" totally unlike the biochemical broth on offer in Viking's GEX café.

The Antarctica 500 sample, later research showed, was also home to extremophiles but of a variety that are best suited to life in frozen conditions. These bugs thrive at temperatures as low as -20°C.[2] Very often they also inhabit tiny pores and cracks inside rocks. Some get their energy from photosynthesis, while others live off inorganic nutrients (mineral soup). Still others, along with the great majority of extremophiles on Earth, dine on whatever meagre supply of organic molecules happens to be in their surroundings – a kind of "chicken soup light."

Extremophiles burst on to the scientific scene in 1977 with the discovery of never-before-seen organisms living right next to boiling hot hydrothermal vents at the bottom of the ocean. Astonishingly, vibrant ecosystems were found thriving in complete darkness fed only by chemicals gushing out in the scalding vent fluids. In the decades since that discovery, extremophiles of every description have turned up in the most unlikely places from the Dry Valleys of Antarctica to the toxic, acidic wastes draining from old mines.

Oyama and his Viking colleagues had no reason at the time to suspect that such alien creatures existed right under their noses. They had no way of knowing there were microbes on Earth so well adapted to life in (by human standards) severely nutrient-deprived conditions that they'd be overwhelmed by a nutrient cocktail that is too rich – much as a starving person would get an upset stomach from a meal of pork ribs and gravy. The Geyserville and Antarctica 500 soil samples *did* contain life, but it was poisoned by the very food used in the equipment designed to detect them.

# 2

# Aliens in my Soup?

On 20 August 1975, atop a Titan III rocket, Viking 1 roared into the blue of a Floridian sky. Two and a half weeks later, it was followed by Viking 2. Both spacecraft spent ten months crossing the interplanetary void between Earth and Mars, covering some 700 million kilometres, before arriving safely at the Red Planet.

On 20 July 1976, after a month in orbit, the Viking 1 lander separated from the orbiter and made planetfall at 11:56:06 UT in the western part of Chryse Planitia (Plain of Gold). Lying some 20 degrees north of the equator, this is a low region into which several ancient channels appear to have drained. Less than two months later, on September 3, the Viking 2 lander touched down about 200 kilometres west of the crater Mie in Utopia Planitia, 28 degrees further north and nearly 180 degrees away from the site of its sister craft. The Utopia region, part of the vast plains that occupy much of the planet's northern hemisphere, is volcanic terrain. With the safe descent of the Viking landers, one of humankind's most extraordinary – and controversial – adventures was about to begin (see figure 3).

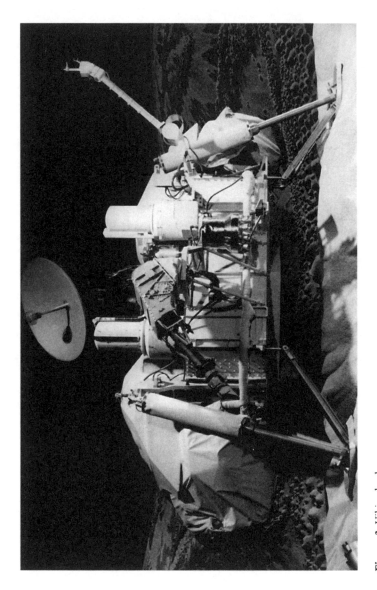

**Figure 3** Viking lander

## Pictures and preparations

**Figure 4** The Martian landscape as imaged by the Viking 1 lander

Twenty-five seconds after settling onto the red sands of Mars, the Viking 1 lander beamed back its first image to Earth – an unprepossessing, black-and-white snapshot of one of its footpads and some nearby stones. Within a day the first colour photo was in, showing a landscape much like the desert of Arizona. An impressive view out to the horizon revealed the red and orange boulder-strewn plain of Chryse Planitia. Weather reports started streaming their way back home: conditions at the Viking 1 site were clear, cold, and uniform (see figure 4), though the permanent haze of dust produced an awe-inspiring pink sunrise and sunset.

Not surprisingly, some technical glitches cropped up during the first few days on the surface. Viking 1's seismometer remained stuck in its protective cage after a pin-pulling device failed to detonate on landing (the same instrument on Viking 2 later deployed without a problem). The spacecraft's UHF transmitter switched to a low power setting before self-correcting back to its normal 30-watt mode within a day or so. More seriously, the sampler arm, whose task was to deliver soil to the biology experiments, the Gas Chromatograph Mass Spectrometer, and the X-ray fluorescence spectrometer, jammed. Technicians feared it might be an electronic problem, and any delay in fixing it would have had an impact on the soil acquisition sequence, scheduled to begin on sol 8. (One sol is a Martian day, which is 40 minutes longer than an Earth day.) Fortunately, the solution proved satisfyingly simple. Engineers realised the arm might be stuck because a locking pin, which was part of the shroud latching system, hadn't dropped free. Sure enough, extending the boom of the sample arm did the trick, allowing the pin to fall to the ground in front of the craft, where onboard cameras could see it.

Everything was now ready for the main business of the mission to begin. Packed into a space smaller than a microwave oven, Viking's biology payload held the hopes and dreams of a race that had long wondered if it was alone in the universe. With the preliminary testing of the spacecraft done, Viking's biology team sent up commands for the life-detection gear to swing into action. Meanwhile the media on Earth were busily hyping up popular expectations that Viking would "prove," once and for all, whether there was life on the Red Planet.

## Positive signs

The first soil samples were scooped up, as planned, on sol 8 (28 July 1976). Four samples were dug, the first being placed into the biology instrument distributor assembly, the next two into the Gas Chromatograph Mass Spectrometer processor, and the fourth into the funnel of the X-ray fluorescence spectrometer. Three days later Viking mission manager Jim Martin reported at a news briefing that biology data had begun to roll in.

From the outset, the messages sent back by the Viking probes were extraordinary – and deeply puzzling. Vance Oyama's Gas Exchange (GEX) experiment on Viking 1 was the first to return data, having begun a cycle in "humid mode," in which the soil was exposed to water vapour alone. A rapid outpouring of oxygen was detected from the sample during the early stages of incubation.

Gil Levin's Labeled Release (LR) experiment reported back next with news of a similar rush of radioactive carbon dioxide when its sample was wetted with radioactive nutrients. Such high levels of activity were surprising, because they were comparable with the results of Labeled Release tests conducted on Earth; true, they were at the lower end of the terrestrial measurements but given that Mars is much drier and colder than Earth, the activity was much greater than most scientists had expected. Something remarkable was happening in the Martian soil – although exactly what was far from clear. The rapid outflow of gas in the Labeled Release experiment slowed down drastically after only 70 hours. If Earth microbes had been in the sample, the activity would have been less pronounced and more drawn out, continuing for perhaps a week longer.

The world's press seized on these early sensational results as virtual proof of life. But the Viking scientists remained cautious. Although living organisms could have been responsible for the dramatic changes seen in the Gas Exchange and Labeled Release experiments, the speed of the reactions was suspiciously like that of chemical processes. Compounds that are highly oxidising – in other words, very effective at transferring oxygen, or more generally, taking away electrons from other substances, and therefore unusually reactive – might, it was suggested, explain the high rate of climb of the gas emissions. Among the chief suspects were superoxides (chemicals that contain negatively charged oxygen in the form of $O_2^-$) and hydrogen peroxide, familiar as a bleach. But the oxidising chemical theory seemed hard to square with the duration of the gas release. As biology team leader Harold Klein remarked early on, though some of the observed activity was probably chemical in origin, there could very well be a life component as well. The three-day period until the gas stopped flowing in the Labeled Release experiment was tantalisingly intermediate between what would be expected with chemistry and biology, and the results of later tests, as we'll see, made the puzzle even greater.

First results from Norman Horowitz's Pyrolytic Release (PR) experiment soon followed and proved no less intriguing. They showed beyond doubt that there'd been some incorporation of radioactively-labelled carbon dioxide and carbon monoxide into organic molecules – just what would be expected from the metabolism of microbes. Instead of a radioactivity count of 15 per minute, which was the value predicted if no gas from the experimental atmosphere were assimilated, the level was 95.9 counts per minute. This value was on a par with what would be achieved by the sparse life in soil from the dry Antarctic deserts and showed clearly that organic synthesis was going on.

Taken as a whole, the early Viking data were perplexing. Mars was giving definite signs of life, but these seemed to be mixed up with some sort of exotic chemical reactions.

Long before Viking left the launch pad, mission scientists had agreed on a key criterion: a positive result from *any one* of the three biology experiments onboard the spacecraft would signify that life had been found. Yet now, with the media spotlight on the team and one of the most momentous calls in human history needing to be made – whether or not we are alone in the universe – the situation no longer seemed so clear cut.

## Through a gas darkly

The confusion deepened when a second shot of nutrients was injected into the Labeled Release experiment on sol 17 (6 August). A fresh surge of radioactive carbon dioxide would have pointed strongly to a biological cause of the reaction. Instead, what happened was completely unexpected. After an initial brief release of labelled carbon dioxide, the amount of gas given off fell to almost two-thirds the previous levels. Researchers were baffled. Some suggested that the gas had been incorporated into microorganisms, others that the equipment had sprung a leak. A few days after the second nutrient injection, the counts of radioactive carbon dioxide started to edge up and were still rising on sol 22 when the incubation period of the experiment's first cycle ended.

There was a palpable sense of excitement among scientists at this stage that life on Mars was creeping closer. The first batch of experiments in which the soil was sterilised before incubation would, it was hoped, give a clearer picture of what was happening. If Martian microbes were responsible for at least some of the activity seen in the first active round of experiments, then this activity ought to be curtailed by killing the bugs in advance.

That's exactly what was seen in the Labeled Release and Pyrolitic Release control experiments. (A control experiment is one designed to show that the factor being tested is actually responsible for the effect observed.) In both the LR and PR control runs, the soil samples were preheated to sterilising temperatures for three hours. The rapid release of gas that marked the active runs of the LR experiment plummeted tenfold, while organic synthesis in the PR control run dropped 85%. "Even our chemically oriented sceptics are excited about these results," said Harold Klein. Head of the PR experiment team, Norman Horowitz, who was later to emerge as an arch sceptic made this key point: "If we had these results in a laboratory we would have concluded that we had a weak positive signal – weak but positive." Yet he cautioned: "Since the signal comes from Mars, which is an entirely different world and one we don't understand yet, we have to be very careful in how we interpret these numbers."

At this stage of the Viking investigation, two facts stood out. First, Mars had at least one chemical on its surface that was extraordinarily reactive. No one on the Viking team had any doubt about that. Second, some of the results were hard to explain without invoking life. Take, for instance, the control runs of the Labeled Release and Pyrolytic Release experiments, which were particularly interesting because they complimented each other. The LR experiment showed that something present only in unsterilised soil could oxidise a nutrient soup to carbon dioxide. The PR experiment, on the other hand, showed that something present only in unsterilised soil could *reduce* (take oxygen away from) carbon dioxide and carbon monoxide in an atmosphere above the soil. Anyone bent on finding a chemical explanation for this double whammy had to conjure up hypothetical reactions never seen naturally on Earth and bordering on the inexplicable. Highly oxidising chemicals in the soil, which broke down when heated, could partly explain the LR data and some other results, but a net reducing soil was needed to make sense of the PR measurements. Only an exotic effect, beyond the realm of known chemistry, could be squared with both sets of results. On the other hand, microbes could easily produce both types of reaction –

as when a plant photosynthesises carbon compounds from carbon dioxide in the air (a reducing reaction), and produces carbon dioxide when it turns food into energy (an oxidising reaction).

The first active run of the Pyrolytic Release experiment also made a powerful case for a blend of high-energy chemistry and life. The measured activity of the sample was hard to fathom without a chemical component, but the unmistakable synthesis of organic material from inorganic building blocks looked authentically biological. In fact, Klein believed at the time that it offered the most convincing argument for life. Only Horowitz, who later became the Viking project's most vociferous opponent of Martian life theories, played down the findings, which ironically came from his own experiment.

Another key observation was made by the Gas Exchange experiment. Not only was oxygen released from the soil in the first active GEX run, so too was nitrogen – an element crucial to life as we know it. Before Viking arrived, a major unknown clouding the biological prospects for Mars was the availability of nitrogen, given that there are hardly any nitrogen-containing minerals. The discovery of nitrogen, therefore, was a huge boost to the hopes for finding life. The nitrogen release itself, however, was open to both a chemical and a biological interpretation. If inorganic, it could have been due to water vapour escaping from the soil and carrying dissolved nitrogen with it; alternatively, it could have come from decomposing microbes, killed after they were inundated with water. Other Viking experiments, in which an uptake of nitrogen was observed, were similarly ambiguous: the explanation could have been chemical adsorption (a process whereby gas molecules stick to the surface of a solid) or biological fixation (as in the case of many Earth microbes which extract nitrogen gas from the atmosphere).

## Life – but no bodies

By the middle of August 1976, all three biology experiments on the Viking 1 lander had yielded provocative results, and a similar set of findings (with a few twists) would soon come from the identical equipment on Viking 2 which was about to touch down. Mars had thrown scientists a huge curve ball with its wildly reactive surface. As mission director Tom Young put it: "We did not properly comprehend how complex the Martian problem was." But for all the puzzles and uncertainties posed by the data, it was hard to avoid the suspicion that Mars looked a little livelier each day.

That was about to change. One simple, shocking piece of news from the Red Planet would abruptly shift mainstream scientific opinion away from the view that Viking might have found life. It came from an instrument whose task was to detect and analyse organic compounds (those containing both carbon and hydrogen) – the Gas Chromatograph Mass Spectrometer. Given the uncertain verdict of the biology experiments, the GCMS took on the role of an appeals court. If there were microbes dead or alive in the Martian soil then there had to be organic compounds.

Connected to the GCMS was a small oven into which a soil sample could be placed. The sample was heated in steps to various temperatures up to 500°C to vaporise or thermally break down any organic substances. The gas chromatograph then separated out the various chemical components produced by the baking process, before passing them on to the mass spectrometer to have their charge-to-mass ratios determined and thus, indirectly, their molecular makeup. A test version of the instrument, identical to the flight units, was used to analyse a number of different soil and rock samples on Earth before and after the Viking missions to evaluate its performance. The scientists in the GCMS project, led by principal investigator Klaus Biemann of the Massachusetts Institute of

Technology, reported that it identified a wide variety of organics in meteorite samples and Antarctic soils in concentrations as low as a few parts per billion. However, this claim applied to gases given off when the sample was heated to 500°C – a temperature not high enough to release gas from some potential organic compounds. Consequently, the limit didn't apply to the total amount of organics that might be present in the sample.

On sol 17 (6 August), the Viking 1 GCMS carried out its first analysis of Martian soil. About 300 mass spectra were beamed back to Earth, showing the various compounds the instrument had found in the sample. None were organic. That wasn't too surprising, as the sample had only been heated to 200°C, which was below the temperature at which organics were expected to break down into small enough fragments to be detected. But what happened next was devastating.

A second run of the GCMS took place on sol 23 (12 August) following reheating of the first sample to a maximum of 500°C. Once again, the GCMS failed to catch even the faintest whiff of organic material. So, too, did a subsequent analysis at the Viking 1 site, and several others at the Viking 2 location. To within the detection limits of the GCMS (a factor to be much debated over the coming years) there were absolutely no carbon-hydrogen-bearing molecules in the Martian surface soil at either site.

This seemed hard to believe. Whether there was life or not on Mars it had been known for many years that some meteorites harbour a slew of organic compounds. Such rocks must have peppered Mars over the aeons, just as they have the Earth. Two decades after the Vikings landed, as we'll see in the next chapter, it came to light that rocks from the Red Planet have actually found their way to Earth and these chunks of the Martian crust carry a treasure trove of organics. What's more the Pyrolytic Release experiment had detected the synthesis of carbon-hydrogen compounds, indicating either the presence of life itself or of chemicals that can easily give rise to substances found in living organisms. Something didn't add up, and a number of Viking scientists questioned what might be behind the negative results of the GCMS at both landing sites.

The team in charge of acquiring samples suggested that the GCMS processor might not actually have received any soil. This was possible because there was no fool-proof way on the instrument of telling whether a sample was in the test cell. When the GCMS instrument was purged, it was hoped to see the expelled soil on the ground but, in the event, the camera view was obscured by the sampling arm. The fact is, there was never any direct, incontrovertible evidence that full or even partial samples were received. The conclusion that samples had been received was based solely on the detection of carbon dioxide and water vapour – compounds which are common in the Martian atmosphere and would have been picked up even in the absence of a soil sample.

Gil Levin and his assistant on the Labeled Release experiment, Patricia Straat, brought up another issue, concerning the sensitivity of the GCMS. They pointed out that their LR apparatus was about a million times more sensitive than the GCMS. Wasn't it possible, they asked, that the GCMS was simply blind to the small amount of organics that would be involved if the LR results were due to microorganisms? Biemann acknowledged that his equipment needed about a billion cells per gram of soil in order to detect organics. But he assured Levin and Straat that there'd be plenty of dead cells from the baking process at 500°C to provide that much raw material. As we'll see later, Biemann's claims of a three-parts-per-billion sensitivity would eventually come under close scrutiny – but only long after most scientists had lost their taste for Viking biology.

It took a few months for all the bad news from the Gas Chromatograph Mass Spectrometers on both Vikings to build up and for the objections of dissenters such as Levin to be put to bed.

One of the last of the fence-sitters to come to terms with the failure of the GCMS to find organics was Viking Project Scientist Jerry Soffen. At first he argued with Tom Young that the test cells must have been empty because there *had* to be organic matter of some sort in the soil. But eventually he was persuaded and was heard to mutter one day, as he walked away from where the data were being analysed, "That's the ball game. No organics on Mars, no life on Mars." Even as the twin Vikings pursued their investigations, NASA announced to the world that the spacecraft had drawn a blank in their life quest: the activity seen in the Martian soil could be explained purely in chemical terms.

## Divergent paths

Almost everyone on the project, as well as the wider scientific community, jumped on to the GCMS bandwagon and took up the mantra "no organics, no life." Reinforcing the GCMS results were those of the Gas Exchange experiment, which could clearly be best explained in terms of chemical activity. The biological case, strongly supported by some of the Labeled Release and Pyrolytic Release measurements, was increasingly dismissed, as too was the fact that the Gas Exchange experimental conditions would likely be hostile to any life on Mars, as pointed out much earlier by Wolf Vishniac.

As the months went by, a clear division opened up among the Viking scientists. Harold Klein and most of the other researchers involved in the life-seeking mission concluded that the results of the experiments, taken as a whole, could best be explained by chemical effects alone. The one major dissenter was Gil Levin who argued forcefully against the general opinion that whatever his Labeled Release experiment had detected it wasn't microbial life.

Klein urged Levin to keep quiet about his suspicions, and tended to shunt him out of the limelight at press conferences. Mission

Director, Jim Martin took a different tack and told him: "Damn it, Gil, why don't you just stand up and say you detected life."

From the outset, Levin was the oddball of the group because of his less academic, more hands-on background. His path to becoming a principal investigator on the Viking mission had been unusual. Prior to his NASA engagement he'd worked as a sanitation engineer for several state health departments, concerned with microbial contamination of drinking and swimming water. He invented a detection method called radiorespirometry, which involved adding small doses of radioactive nutrients to the biochemical soups. Organisms couldn't distinguish between normal and radioactive nutrients and any consumption could easily be measured with a radioactivity counter – to an astonishing level of sensitivity (about 10 bacterial cells per sample).

Levin's association with the space agency began in 1958 when he accompanied his wife to a Christmas party and there met the first NASA administrator, Keith Glennan. During a conversation, Levin spoke about his research and Glennan suggested that he send a proposal to Clark Randt, head of the new NASA biology program. Levin's proposal to develop the radiorespirometry experiment for use on Mars was eventually selected, in the face of fierce competition, and renamed the Labeled Release experiment.

Even in the early days of Viking, as the mission started to take shape, the potential for disharmony among the project scientists was clear. Levin was an engineer by training and disposition, and was very upbeat about the prospects for life on Mars. Norman Horowitz, like most of the others on the Viking science team, was a career academic, and of everyone involved in the mission, the least hopeful of finding life. It's hardly surprising that Levin and Horowitz eventually found themselves at loggerheads.

As the Viking landers moved forward with their investigations, Levin and his co-worker Patricia Straat insisted that the Labeled

Release results were in good accord with a biological interpretation. Central to their claim was the remarkably uniform production of gas from the LR nutrient when it was added to soil samples at both lander sites, and, more importantly, the biologically-consistent responses from the whole range of heat-treated control samples.[3]

Exposing a duplicate sample to the one that gave a positive response to 160°C for three hours rendered it inactive. This satisfied a pre-mission criterion for a life signal, argued Levin, because any likely chemical reagents would have survived such heating and given another positive response. What's more, further tests showed that the active agent in the soil was destroyed at 51°C, reduced by 70% at 46°C, and, most tellingly, eliminated altogether after standing three months in the dark inside the sample container held at 7° to 10°C.

The samples did, however, keep their activity for up to several Martian days in the sample test chamber held at approximately 10°C before testing. No chemical oxidant, among the many proposed, said Levin, could duplicate this thermal sensitivity profile.

Despite these arguments, the consensus within the Viking science team, and within NASA as a whole, shifted relentlessly toward a non-biological explanation. Theories about what inorganic compounds and reactions might have caused the Viking results became the order of the day. Most invoked some kind of very strong oxidant, or combination of oxidants, which would react with water to produce oxygen and hydrogen, and with nutrients to generate carbon dioxide. This notion, that powerful oxidisers were at work, fed the suspicion that Mars was sterile because such compounds would, it was assumed, be inimical to life as we know it.

Howard Klein freely acknowledged that there was room for doubt: each of the Viking biology experiments operated under conditions that weren't the same as those actually found on Mars. "While we have obtained significant and fascinating data in the

Martian experiments," he said, "we may not have hit upon the proper conditions to elicit evidence of Martian metabolism."[1] His statement mirrored the realisation that the Viking life detection experiments were carried out before we had a proper handle on the Martian environment and therefore a solid basis on which to interpret the results.

Klein concluded that while some of the results fitted a biological interpretation, most were hard to reconcile with life as we know it. Interestingly, the experiment that he thought made the best case for life was the Pyrolytic Release experiment. "An explanation," he said, "for the apparent small synthesis of organic matter in the PR experiment remains obscure." However, the principal investigator of the PR experiment, Norman Horowitz, was never in favour of Martian biology. In fact, from the outset, Horowitz had such strong doubts about finding anything alive on Mars that on several occasions other members of the team wondered aloud why he had remained with the group.[1]

In 1978 Klein published his, and effectively NASA's, definitive word on the subject. He weighed the various chemical and biological explanations on offer and concluded that, while some of the data were consistent with a biological interpretation, most were better in tune with a reactive chemical scenario.[4] Most scientists rallied around Klein's assessment and started theorising about which strong oxidiser might best explain the Viking results. Hydrogen peroxide was an early favourite, but no one could come up with a mechanism for making enough of it to match the responses of the biology experiments, particularly the Labeled Release experiment. It also wasn't clear how this compound could be stable enough on the Martian surface to accumulate amid the battering of intense ultraviolet radiation and other destructive forces. If hydrogen peroxide were the active agent, it would have to be wrapped up and protected somehow in chemical complexes with minerals in the soil.

Although the mysterious oxidant couldn't be pinned down, sci-entific support for the interpretation that all observations could solely be explained by chemical reactivity remained strong, and Levin and Straat were increasingly sidelined. Meanwhile NASA was so convinced that Mars was dead that the Viking biology program was swiftly wrapped up and its principal investigators left to look for new jobs. Extraterrestrial life research at NASA continued only through the poorly-funded Exobiology Branch until the Astrobiology Institute was set up in 1998. Ironically, the two main events that spurred the formation of the Astrobiology Institute were the possible detection of fossilised life in a Martian meteorite (as described in the next chapter) and the surprising discovery that life thrived in a variety of extreme environments on Earth.

## Oxidants? What oxidants?

The idea that the Viking results were caused by highly oxidising compounds born of the interaction between ultraviolet radiation and the Martian soil became the paradigm of choice. This was despite the fact that the theory failed to explain a number of key observations. Most obviously, it left unanswered the question of why the Pyrolytic Release experiment recorded the synthesis of organic material.

Among other things, it also didn't address a curious outcome of one of the Labeled Release experiments on Viking 2. In this exper-iment, the reactivity of soil samples taken from in the open and under a rock, nicknamed Notched Rock, was compared. It turned out that the soil which had been sheltered by the rock, and therefore in the dark, was almost as reactive as soil that had been lit, and far more reactive than samples that had been heated to sterilising tem-peratures – or even to a mere 50°C. Although light was evidently a factor in boosting the reaction rate, just as clearly some of the reaction was still able to take place in darkness. This was hugely significant because it implied that some of the reactivity seen in the Pyrolytic Release and Labeled Release experiments couldn't be due to an oxidiser that forms from radiation interacting with the Martian soil. The patch sampled under the rock at the Viking 2 site had surely not been illuminated for millions, and perhaps even tens or hundreds of millions of years. Yet the soil gave a positive response to the nutrients offered to it in the LR run. That was a puzzle for anyone favouring a purely chemical hypothesis because it's hard to envision how any highly oxidising compound – one that can supposedly split apart water into oxygen and hydrogen (as in the GEX experiment) – could come about without the direct influence of strong radiation. The surprising reactivity of the under-rock soil *could*, however, be explained by the presence of microbes which relied not on sunlight but on a readymade supply of organics.

The fact is that no chemical model so far devised adequately mimics the Viking results. What's more, no suitable oxidant, including hydrogen peroxide, has been detected at the levels required by any subsequent mission to the Red Planet. The bit of inorganically-produced hydrogen peroxide that *is* present on Mars occurs at levels less than one per cent of those needed to explain the Viking responses.

Most damning of all, the same type of Gas Chromatograph Mass Spectrometer that failed to detect any organic matter on Mars also reported a sample of Antarctic soil to be sterile, even though a later wet chemistry analysis demonstrated the presence of organic matter and the likely presence of microorganisms in a similar sample. This was the very instrument used to settle the dispute between biological and chemical interpretations of the Labeled Release, Pyrolytic Release, and Gas Exchange data. In 2000, the Viking GCMS came under fresh scrutiny when chemist Stephen Benner, then professor at the University of Florida in Gainesville and now director of the

Westheimer Institute of Science and Technology, pointed out that it would have been unable to detect certain organic compounds potentially critical to a life investigation.[5] More recent and devastating was the blow to the instrument's credibility dealt by a group of scientists led by Rafael Navarro-Gonzalez of the National Autonomous University of Mexico. This group reported that the sensitivity of the Viking GCMS was several orders of magnitude lower than originally thought.[6]

At the same time, some of the key arguments levelled against a life interpretation of Viking's findings have been undermined. One of these criticisms is that life needs water and there was no evidence of water at the Viking sites. However, frost *was* observed at the Viking sites and extensive regions had already been found on Mars by the Mariner 9 orbiter where the surface pressure exceeded the triple-point pressure (the temperature and pressure at which a substance can exist simultaneously in its solid, liquid, and gaseous states) of water thus allowing water to exist in the liquid phase. Furthermore, Viking data indicated that the temperature of the top several millimetres of soil beneath the Viking 2 lander sampling had risen to 0°C (where ice liquefies) and remained there for at least several minutes.[7] Also, the advancing science of extremophiles – life in extreme environments – has shown that many organisms can survive being dried out in a dormant state for extraordinarily long periods, perhaps even hundreds of millions of years.

Harold Klein insisted that no known Earth organism had been shown capable of reproducing all the Viking results. However, any Martian microbes would inevitably have adapted over many millions of years to the conditions found on the planet today. The critical question then becomes: what are the ultimate limits of life under the environmental conditions to which a soil is exposed? For example, we know that the Martian environment is poor in nutrients. One strategy that terrestrial microbes use to deal with nutrient-poor environments is to become small. Bacteria in a growing state typically have a cell diameter of about one micron (millionth of a metre), but many bacteria can shrink to one-tenth of this size or less when faced with starvation or other extreme stresses. Other survival strategies used by microscopic life are to become highly efficient in metabolising the few nutrients on tap, or to go dormant and then reproduce quickly when food becomes available.[8]

## Life – but not as we know it

The only spacecraft ever to search for life on another world was built to look for the kind of microbes with which scientists were most familiar in the 1970s. Viking's experiments supplied a specific, narrow range of environments and nutrients that, in pre-flight tests, had proved attractive to common-or-garden Earth bugs. These experiments were also designed in the expectation (known as the Oyama model) that if life did exist on Mars, it would be spread over the whole planet at least as densely as bacteria occur in challenging locations such as Antarctica.

A lot has been learned both about the limits of life on Earth and about conditions on Mars in the three decades since Viking. An extraordinary variety of extremophiles has come to light, vastly broadening the scope of potentially habitable places on other worlds, and thanks to a number of advanced robotic probes in recent years we've added terabytes to our library of data on the surface, subsurface, and atmospheric environments of Mars.

Yet, for all this new knowledge, there's still a tendency to think of life from a terrestrial perspective. Nowhere is this more evident than in the persistent arguments put forward to account for the Viking results in chemical rather than biological terms.

A very different picture emerges, however, if we free ourselves from the shackles of geocentrism. Why on Earth – or, rather, on

Mars – should organisms on the Red Planet exploit the same bio-chemistry and evoke the same biological responses as terrestrial microbes? Whether Mars came up with its own life from scratch, or received an early biological gift from Earth via meteorites, adaptation and evolution under alien conditions would surely have given rise to species today that have alien biosignatures.

It was with such thoughts in mind that Joop Houtkooper, of the University of Giessen in Germany, came up with a novel idea of life on the fourth planet. Houtkooper was initially inspired by an essay by the physicist Freeman Dyson called "Warm-Blooded Plants and Freeze-Dried Fish" and first presented his concept of Martian fauna at the Bioastronomy 2004 conference held, appropriately enough, in Iceland. Shortly after, Schulze-Makuch (one of the co-authors of this book) teamed up with Houtkooper to develop this idea further.

Put simply, the two scientists envisioned microbes on Mars that use not plain water inside their cells (as Earth life does) but a mixture of water and hydrogen peroxide.[9] The phrase "peroxide blonde" is a familiar one. Hydrogen peroxide ($H_2O_2$) is a potent chemical that's just the job if you fancy bleaching your hair, but isn't the sort of stuff you'd want sloshing around inside your body. At the same time, it isn't entirely unknown among creatures on Earth. Some bacteria produce it and even use it in their metabolism, while keeping its reactivity in check with a chemical stabiliser. There's also the remarkable case of the Bombardier beetle (see figure 5), which produces a solution of 25% hydrogen peroxide in water and sprays it on to any unfortunates that it considers a threat.

Given that some terrestrial life forms have evolved to be able to harbour high concentrations of hydrogen peroxide, it's by no means far-fetched to speculate that microbes on Mars might have made it an integral part of their biochemistry. An intracellular cocktail of hydrogen peroxide and water would offer a number of benefits to organisms in the cold, dry Martian environment. The freezing point

**Figure 5** The bombardier beetle is probably the most spectacular example of the use of hydrogen peroxide in life on Earth. Although the hydrogen peroxide is not used for metabolism in this organism, it shows that high concentrations of hydrogen peroxide and organic tissue are compatible.

of a hydrogen peroxide solution can be as low as −56.5°C (depending on the peroxide concentration); below this temperature, the solution becomes firm but doesn't form cell-destroying crystals, as water ice does. Hydrogen peroxide is hygroscopic, which means that it attracts water vapour from the atmosphere − a valuable trait on a planet where liquid water is scarce. And, finally, peroxide offers a rich source of oxygen to power cellular activity.

On Earth, there are good reasons why life doesn't generally embrace the potential benefits of hydrogen peroxide. Earth is two-thirds covered by water and so not the best place for a hygroscopic and reactive compound to prove its merits. If organisms here need a form of antifreeze they tend to use salts, which are highly soluble in water and mimic the chemistry of Earth's oceanic environment. Early terrestrial organisms were exposed to salty ocean water and learned to adapt to salt in high concentrations, and further use it as antifreeze in cold places such as mountainous regions and the Arctic.

On Mars, liquid water was never as plentiful as on Earth. Even during the early, wet phase of Martian history, its seas and lakes were not long-lived in geological terms. Organisms on the fourth planet could have adapted to use the properties of hydrogen peroxide to their advantage, especially as the planet became increasingly dry. Over time, the hygroscopic character of this potent chemical, a handicap on the wet Earth, would become an advantage on our desiccated outer neighbour.

## The Viking mystery solved?

The possible existence of powerful oxidising chemicals, including peroxides, had been the most popular conventional explanation for Viking's surprising observations, and also a reason to assume that life might be impossible on or near the surface of Mars today. What this new theory does is turn the argument on its head by proposing that living Martians actually make use of hydrogen peroxide in running their cellular machinery. If they do, then, remarkably, it would explain all the most puzzling results of the Viking biology experiments.

The peroxide–water hypothesis addresses all the major results from the life detection experiments plus the non-detection of organics by the Gas Chromatograph Mass Spectrometer. In particular, it offers a good explanation for problematic issues that have haunted the mission since the data were first received.

The Viking GCMS heated the Martian soil samples to temperatures of several hundred degrees Celsius to analyse them for organic compounds. Such heating would kill any organisms. Being a powerful oxidant, hydrogen peroxide, when released from dying cells, would sharply lower the amount of organic matter in its surroundings and produce carbon dioxide instead (which was measured by the GCMS). Combined with the fact that the GCMS was much less sensitive than originally thought, this would explain why the instrument didn't detect any organics on the Martian surface.

The new hypothesis would explain the nature of the mysterious oxidant on Mars, since the hydrogen peroxide in the internal cellular fluid is part of the very biochemistry of the Martian organisms.

The most puzzling result from the Gas Exchange experiment was the enormous release of oxygen. This can be interpreted in two ways: as the result of an energy-producing metabolism, or more likely, upon humidification, as due to the decomposition of Martian organisms. After death, the microbes' internal peroxide would disintegrate under Mars' surface conditions into oxygen and water, catalysed by inorganic oxides in the soil.

The Labeled Release experiment, in which samples of Martian soil were exposed to water and a nutrient source including radio-labelled carbon, showed rapid production of radio-labelled carbon dioxide which then levelled off. The initial increase would have been due to metabolism by peroxide-containing organisms, and the later levelling off to the organisms dying from exposure to the environmental conditions. The possibility of die-off in the LR experiment had been mentioned by Levin earlier,[10] but he couldn't pinpoint the reason for it. The peroxide-water hypothesis explains why the experimental conditions would have been fatal: microbes using this mixture would either "drown" (in scientific terms, suffer hyperhydration), or they would burst, due to water absorption by the hygroscopic peroxide, if suddenly exposed to water.

The possibility that the tests killed the very organisms the Viking team was looking for is also consistent with the results of the Pyrolytic Release experiment, in which radio-labelled carbon dioxide was converted to organic compounds by samples of Martian soil. Of the seven tests done, three showed significant production of organic substances and one showed a much higher production prior to death. The variation could be due to patchy distribution of microbes, but perhaps most intriguing is that the sample with the lowest production (Utopia 2) – lower even than the control – was the only one that had been treated with liquid water.

Only on Mars can the peroxide–water hypothesis of life be properly tested. Either we can use data sent back by current and future missions, and look for results consistent with the hypothesis, or we can design a new mission that tackles the issue head-on. Current missions are not particularly helpful in this regard because they're not designed to search for oxidants or life. All they can do is support the case that Mars was, and potentially still is, habitable, based on the availability of water. However, future missions, such as NASA's Mars Science Laboratory and the

European Space Agency's ExoMars, may shed more useful light on the subject.

Seen through the eyes of twenty-first century astrobiology, the Viking results make a strong case for active biology on Mars. But they are not alone. Recent discoveries, wholly unexpected, have greatly strengthened the claim that life existed, and still exists today, on the Red Planet.

# 3

# Rock Star

Chunks of rock rarely make world headlines. But that's exactly what happened in August 1996 following a NASA press conference hosted by the White House. David McKay of the Johnson Space Center and his research colleagues claimed they had several good reasons to suspect that a potato-sized meteorite from Mars, called ALH 84001, had once harboured microscopic organisms. In the words of President Bill Clinton: "Today, rock 84001 speaks to us across all those billions of years and millions of miles. It speaks of the possibility of life."

## Claims and controversies

The press conference quickly turned into a media frenzy. Hundreds of journalists jostled to get closer to NASA's Administrator, Dan Goldin, and the scientists who flanked him. After Goldin's introduction, the team members gave their well-rehearsed presentation. A video explained how the meteorite had made its journey from the Red Planet to Earth. Images were shown of what the research group suggested were fossilised Martian microbes embedded in the rock (see figure 6).

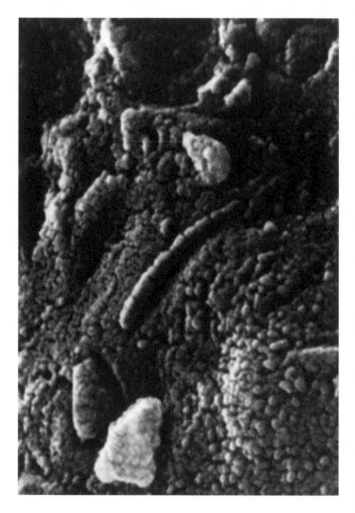

**Figure 6** The image that went around the world. The segmented worm-like centre structure was interpreted by some to be a Martian microbe in meteorite ALH84001

McKay and his colleagues said that a detailed report of their findings would soon appear in the journal *Science*.[11] Their claim of extraterrestrial biology centred on a handful of key observations, which they argued were consistent with an interpretation of life.

The team acknowledged that none of the observations alone clinched the case for past life on Mars. Each, by itself, was open to alternative, non-biological explanations. However, said McKay, taken as whole, particularly in view of the way the structures and chemicals were arranged within the meteorite, the data amounted to powerful evidence for primitive life on early Mars.

Also on the panel of scientists at the press conference was Bill Schopf, a paleobiologist at the University of California, Los Angeles. He'd been invited by Goldin to play the role of sceptic. The NASA Administrator, not a scientist himself, but a business manager and politician, knew that his agency's claim of extraterrestrial life would draw plenty of flak from academic circles. Someone like Schopf, he reasoned, would serve as a much-needed counterweight to the controversial announcement. In his book *Cradle of Life*, Schopf eloquently describes the details of this historic moment and the doubts he expressed at the conference about the meteorite's evidence for life on Mars.

Not surprisingly, given the media's preference for shock and awe, the news headlines that day and the next played up the sensational. Pictures of the wormlike structures touted as possible Martian bugs were splashed across network TV and newspaper front pages alongside breathless commentary.

But the more important analysis and debate, within the scientific community, was only just beginning. The claim for ancient life on Mars, based on ALH 84001, put astrobiology in the public spotlight as never before and, as Goldin had hoped, quickly attracted new government money for the subject. But were those really fossils of

Martian microbes? Was ALH 84001 the smoking gun that pointed to life on our planetary neighbour in the past and, possibly, even today?

## Exo-forensics

The pristine blue ice fields of Antarctica are rich hunting grounds for those in search of meteorites. So few Earthly rocks lie on the surface of this frozen wasteland that anything dark and stone-like is, by default, most probably from space. ALH 84001 earned its name for being the first find in the Allen Hills region of Antarctica in 1984 during that season's National Science Foundation Antarctic search for meteorites.

From a close study of its minerals and internal structure, geologists gleaned some pretty secure facts regarding the history of ALH 84001. It is one of the oldest meteorites ever found. It formed several kilometres below the Martian surface, deep within the congealing crust of the planet about 4.5 billion years ago, shortly after Mars itself coalesced from the solar nebula – the cloud of matter which spun around the infant Sun. Some 3.6 billion years ago, the original material of ALH 84001 was shocked and shattered by one or more asteroid impacts on the surface, at a time when Mars was much warmer and wetter than it is now. Groundwater likely seeped through the fissures and fractures of the rock and filled them with carbonate material. Finally, at a time when our earliest primate ancestors were evolving, another asteroid struck, with such tremendous force that it hurled the future ALH 84001, along with other debris, into space and on to a trajectory that would eventually bring it to Earth. It remained in space until about 13,000 years ago when it was pulled in by Earth's gravity and fell to the ground in Antarctica. There it lay until its discovery in 1984.

Visually, ALH 84001 was the most unusual meteorite found that year, but it would be another nine years before scientists realised just how extraordinary it was: an actual piece of the planet Mars. The telltale signs were in its mineralogy and, most convincingly, the makeup of gas trapped within tiny pores of the rock.[12] The Viking mission and later Mars rovers revealed a very distinctive blend of gases in the Martian atmosphere. Nowhere else in the Solar System has that mixture been found. But it was here, unmistakably, within minute pockets of the meteoritic rock. Even today, the tally of known rocks on Earth with that same Martian signature stands at barely 30.

David McKay and his colleagues took great pains, during handling, to avoid contaminating their precious piece of Mars with anything remotely biological from Earth. Contamination by terrestrial microbes and biochemicals is the biggest problem that faces researchers when looking for traces of alien organics in meteorites. It's the first question that other scientists raise when claims of extraterrestrial life or life-related substances are made based on meteoritic evidence. The NASA team were able to counter the argument that they may have introduced contaminants themselves. But what about the 13,000 years during which ALH 84001 had been lying around before anyone noticed it?

One of the reasons Antarctica is prized as a meteorite hunting ground is that the processes of weathering are so slow and feeble there. ALH 84001 couldn't have been better preserved if it had been kept in a low-temperature freezer for its 13 millennia on Earth. Tests also showed that the organic carbon content of the meteorite increased toward the centre, levelling off at a depth of about 1.2 millimetres below the surface. This trend could be explained purely and simply by the meteorite's passage through Earth's atmosphere. During its fiery fall, ALH 84001's outermost layer would have melted, producing a thin fusion crust more or less devoid of volatile (heat-sensitive) organics. If the organic matter found in the meteorite had come from Earth, it would have been more prevalent on the outside rather than the inside of the rock. Also, the organics in ALH 84001 are in the parts per million range, which is thousands of times higher than the organic content detected in other meteorites found in Antarctica, and an estimated million times higher than could be accounted for by natural contamination.

Evidently, ALH 84001 had suffered virtually no contamination either before or after its discovery. Therefore, the structures, minerals, and organic material within the rock were, beyond reasonable doubt, Martian in origin. This, in itself, was a stunning conclusion, which undermined one of the main arguments against a biological explanation for the Viking experiments. There was, after all, organic material on Mars. But was that material directly related to life? McKay and his team based their claim that it was on a variety of clues.

## Lines of evidence

There were carbonate globules locked away inside the meteorite. The larger of the globules had cores rich in the elements calcium and manganese, surrounded by alternating bands loaded with iron and magnesium. These bands contained the minerals magnetite (a type of iron oxide) and iron sulphide, which speak of two opposing chemical processes – oxidation and reduction – happening very near together. On Earth, the close association of such processes, and its implication of something happening far out of equilibrium, is a strong indicator of biological activity. What's more, it's well known that particles of magnetite and iron sulphide are laid down alongside each other within individual bacterial cells. The magnetite particles in ALH 84001 were strikingly similar to magnetite particles found in some organisms on Earth. This kind of magnetite has also been found in ancient terrestrial limestones (made of calcium carbonate) and been interpreted to be of biological origin.

Most controversial was the claim that the jelly-bean-shaped and threadlike structures in the images represented fossilised microorganisms (see figure 6). The researchers acknowledged that these structures were much smaller than typical microbes on Earth, but they drew attention to a recent report of exceptionally small microbial organisms, named nanobacteria, by Robert Folk from the University of Texas.[13] Interestingly, these are about the same size as the structures in the meteorite. Some of the objects in ALH 84001 were about 380 nanometres long (one nanometre is one millionth of a millimetre) – just big enough to pack in all the cell components needed for a typical Earth microbe to function. However, some of the other structures claimed to be of a biological nature were much smaller, 20 to 170 nanometres long. These, the researchers concluded, might represent fragments or appendages of microbes, similar to the ones observed in samples of basalt rock found near the Columbia River in Washington State, USA.

## Cold reception

While the claim of life in meteorite ALH 84001 created a huge media stir, it faced a stern test in convincing the scientific community. Kathie Thomas-Keprta, a research scientist at Lockheed Martin and a co-investigator in David McKay's group, quoted from Mark Twain at the Lunar and Planetary Science Conference in 2002: "The scientist will never show any kindness for a theory which he did not start himself."

McKay and his team maintained that while none of the lines of evidence by itself clinched the case for Martian life, taken as a whole they pointed to life as the most straightforward explanation. The group's position was summed up by Everett Gibson, from the Johnson Space Center and co-author of the original paper on the possible relic biogenic activity in the Martian meteorite: "We think there's a small percentage of the processes recorded in the rock, 25%, that may represent the end product of biological processes. So you have to sort out what the vast majority of the signatures are, which are the inorganic, normal production of a rock and other processes on Mars. Biology is a small ingredient in the life of a planet. And in the case of Mars, it is only a small component, yet it leaves a trace signature."

Most spectacular of the claimed footprints of Martian life were the fossil-like forms, photos of which were splashed across newspapers and magazines worldwide. But, from the outset, many scientists were highly sceptical that these tiny mineralised shapes had ever been alive. First there was the issue of how anyone could know what a Martian microbe was supposed to look like. We have trouble enough identifying very simple Earth organisms from the fossil record. It was true that the worm-like structures in ALH 84001 bore a superficial resemblance to primitive, one-celled creatures from our own planet, but looks could easily deceive. Second, and more serious, was the problem of their size – the ALH 84001 "fossils" were, for the most part, a lot smaller than typical Earth bacteria.

McKay's group tried to pre-empt the size concern in their original paper by citing the recent discovery of nanobacteria – a class of previously unsuspected super-small organisms living on Earth. Several researchers, in addition to Bob Folk, came forward over the next few years with evidence of mineral-forming nanobacteria in all sorts of unusual places, including human kidney stones and blood.[14] However, the main problem that most biologists had with nanobacteria was that they seemed too small to be able to hold the minimum machinery, including DNA, needed for a cell to function. The National Academies of Sciences convened a workshop called "Size

Limits of Very Small Organisms" to address this issue and concluded that any free-living organism would have to be at least 250 nanometres in diameter. That conclusion was accepted for a while, but has now been shown to be wrong. In 2006, a research team led by Brett Baker from the University of California at Berkeley found three varieties of primitive microbe, each containing DNA and RNA, with diameters of less than 200 nanometres.[15] These minuscule bugs were discovered thriving in slime on the floors of mines, converting iron compounds into acid as part of their daily metabolism. It seems that, after all, there are Earthly creatures as diminutive as many of the fossil-like structures in ALH 84001 and, what's more, these ultra-tiny microbes make a living in a way that would be very handy in the iron-rich environments of Mars.

Yet, despite the newly appreciated size limit on living things, of all the evidence put forward by McKay's team, the least scientifically convincing remains the supposed fossils. Much more interest, in the astrobiological research community, has focused on the intriguing carbonate and magnetite deposits within ALH 84001.

The central issue of the carbonate globules, right from the start, was the temperature at which they were laid down. If further analysis revealed that they had formed at temperatures high enough to sterilise life as we know it, then the biological argument would be seriously undermined. For a number of years, debate about the carbonates swung back and forth as different groups published their various results and opinions. Finally, the question appeared to be resolved when two groups, one led by John Eiler from the California Institute of Technology and the other by Christopher Romanek from the Johnson Space Center, confirmed the presence of at least two chemically distinct carbonates.[16] One type had formed at a high temperature, while the other had been precipitated from low-temperature liquid water – a finding consistent with the biological explanation.

## Magnetite – a pointer to life on Mars?

The fiercest debate erupted over the magnetite grains in ALH 84001. According to McKay's group, these were clear indicators of life on the Red Planet.

Magnetite is an iron oxide mineral that famously makes compass needles point to magnetic north. Tiny magnetite crystals are also used by living things on Earth for orientation purposes. The ability of pigeons, for example, to navigate accurately on their long voyages depends on microscopic magnetite crystals in the birds' brains. Likewise, magnetite crystals enable certain bacteria, called magnetotactic bacteria, to align themselves with Earth's magnetic field (see figure 7). These tiny biological magnetite crystals are extraordinarily pure and have characteristic shapes that distinguish them from chemically produced magnetite.

Intriguingly, the magnetite crystals in ALH 84001, too, are exceptionally pure and distinctive in shape. McKay's group, spearheaded on the magnetite issue by Kathie Thomas-Keprta, studied 594 magnetite crystals from the Mars meteorite and grouped them into three populations based on their outer appearance: 389 were irregularly shaped, 164 were elongated prisms, and 41 were whisker-like.[17] The group identified six properties which, it said, taken together, indicated the work of biology. These pertained to the extraordinary chemical purity of the crystals and their distinct shapes, sizes, and geometry. A quarter of the magnetite crystals in ALH 84001 conformed to all six properties, leaving three-quarters that could be explained by chemical processes.

Thomas-Keprta and her colleagues pointed out that none of the best inorganically or artificially produced magnetites could satisfy all six criteria, despite a $35 billion a year industry trying to produce the best magnetites possible. Also, they said, the association of biological magnetites with chemically-produced ones in ALH 84001

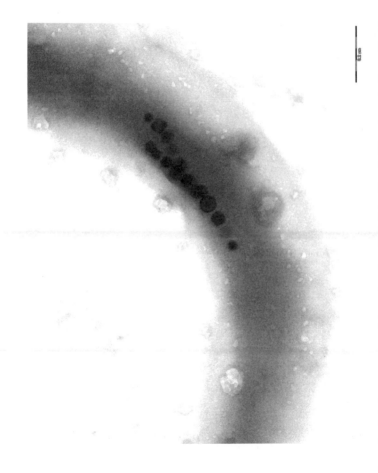

**Figure 7** Magnetite chain in a magnetotactic bacterium from Kelly Lake, Canada. These bacteria align to Earth's magnetic field.

wasn't surprising, because magnetites formed by bacteria on Earth are usually found in environments where chemically produced magnetites are common. The group made an even bolder claim – that the magnetite crystals constitute evidence of the oldest life yet found.[18]

Of special interest was the close similarity between the geometry of the magnetite crystals in ALH 84001 and that of the magnetite found in a particular bacterium called MV-1. This organism gives rise to only one type of magnetite crystal, in the narrow size range of 30 to 120 nanometres. MV-1 is known to produce a single chain of about 12 well-ordered magnetite crystals inside its cell, each encapsulated within a coating or membrane. The chain of magnetite crystals acts like a compass needle and allows the passive alignment of the bacterium along the Earth's magnetic field lines. This type of chain is exactly what Imre Friedmann, a biologist at Florida State University, and Jacek Wierzchos, an expert in microscopy from Spain, discovered in ALH 84001.[19] Such an alignment would be nearly impossible to generate chemically and argues powerfully for a biological origin.

Still, plenty of criticism was levelled at the biological interpretation of the Martian magnetite. A research group led by John Bradley of the Georgia Institute of Technology published its own analysis of ALH 84001's magnetite crystals, according to which some of the magnetite whiskers and platelets had most likely formed at 500 to 800°C – temperatures wholly inconsistent with a life origin.[20] However, it emerged that Bradley's group had focused on magnetites that were due to chemical processes, so that their analysis didn't detract from the claim that other crystals involved biology.

David Barber from the University of Greenwich, London, and Edward Scott from the University of Hawaii suggested that all occurrences of magnetite in ALH 84001 could be accounted for by mineral growth associated with the heat of an asteroid impact.[21] They concluded that biogenic sources shouldn't be invoked for any of the magnetites. But Barber and Scott's claim was undermined by the fact that the heat necessary to decompose the mineral iron carbonate and form magnetite simply wasn't present, and, if it had been, would have erased the magnetic signatures seen in ALH 84001. Also as mentioned earlier, research on the thermal history of the meteorite favoured a low-temperature origin of the carbonates hosting the magnetites.[22]

A balanced perspective on the magnetite issue was offered by a research group led by Benjamin Weiss from the California Institute of Technology. He and his co-workers argued that magnetite crystals in the carbonate from ALH 84001 had a composition and shape indistinguishable from that of known magnetotactic bacteria.[23] They noted that the alignment of magnetites in chains was the strongest argument for a biogenic origin of the magnetite and that no more than 10% of the magnetite in ALH 84001 occurs in chains. They also pointed out that that the magnetite is unusually pure and fine-grained, similar to biologically produced magnetite on Earth, but that it would be difficult to conclusively prove the biological origin of the magnetites due to the low abundance of chains. During the natural preservation process, turning living organisms into fossils, many of the chain structures are naturally disrupted.

## New signs of ancient life in another Martian meteorite?

Ten years after their ground-breaking announcement, at the Lunar and Planetary Science Conference and the Astrobiology Science Conference, both held in the spring of 2006, McKay and his team dropped a new bombshell. They showed images of complex organic matter within microscopic veins in another Martian meteorite, known as Nakhla, which were similar to what they'd seen in ALH 84001.[24]

At the same time, an independent group, led by Martin Fisk of Oregon State University at Corvallis, reported finding tunnel- and borehole-like structures in both basalt rocks from Earth and the Nakhla meteorite from Mars.[25] Fisk and his colleagues pointed out that these tunnel-like structures tested positive for the presence of biological material in the basalt rocks from Earth, and that the tunnel structures in the Nakhla rock were indistinguishable in size, shape, and distribution from the terrestrial samples. Nakhla and one other meteorite with a similar Martian origin, called Lafayette, were found to harbour these intriguing structures.

Suddenly, some earlier pieces of research, which had been almost forgotten, took on a new significance. The scientific spotlight turned to tunnel structures discovered in 3.5-billion-year-old rocks from South Africa. A study led by Harald Furnes of the Department of Earth Science at the University of Bergen, Norway, had uncovered micrometre-scale tubes filled with minerals that suggested submarine microbial activity during Earth's early history.[26] Microbes had etched little fractures into the rock and left behind traces of DNA associated with the tiny tubes in the basalt. The Norwegian researchers also discovered chemically altered, isotopically-light carbon indicative of biological processing – the same type of chemically altered carbon that they found in the Nakhla rock, suggesting a biological origin of the etchings in the Martian meteorite as well.

McKay's group was quick to point out that the organic stuff in Nakhla wasn't only rich in carbon but also in nitrogen, another essential element for life. Interestingly, back in 2001, Everett Gibson and a number of collaborators making the past life on Mars claim, including McKay and Thomas-Keprta, reported in the journal *Precambrian Research* about possible biological features in Nakhla and another Martian meteorite called Shergotty, but this research got little attention at the time. Now, with similar features having been recognised in Earth rocks, the Nakhla and Shergotty observations looked a lot more significant.

Today, the debate rolls on. There's no doubt that the parallels between the micro-tunnels and organic residues discovered in rocks from Earth and those of the Martian meteorites have re-energised the pro-life argument. But alien forensics is a tricky business, especially when working with specimens that are more than a billion years old. Critics of the hypothesis for fossilised life on Mars can still point to unanswered questions, notably about the Nakhla rock. In particular, there are uncertainties hanging over the time when Nakhla formed, and the time and place it fell on Earth.

Opponents of the microbes-in-meteorites claim still have room for manoeuvre. Yet, the evidence from various independent groups is mounting that these ancient specimens do indeed contain evidence for biology that argues persuasively for the presence of past life on Mars. There seems little doubt that if these rocks - ALH 84001, Nakhla, and Shergotty - had been terrestrial in origin, the structures and materials they bear would have been accepted beyond reasonable doubt as evidence of life.

This whole business is reminiscent of the arguments about the outflow channels (ancient riverbeds) on Mars, which many scientists were resistant to admit were carved by liquid water (see figure 8). Only spacecraft which recently verified large amounts of water ice in the shallow subsurface of Mars have swayed opinion in favour of liquid water. Carl Sagan's words come to mind, that the person who makes extraordinary claims needs extraordinary evidence to convince his peers. The question is: what constitutes extraordinary evidence? Can any supposed chemical and biomarkers sway the case, or will only a living Martian organism provide final proof?

It seems that in the case of ALH 84001 and its ilk a stalemate has been reached similar to that of the Viking experiments. There's

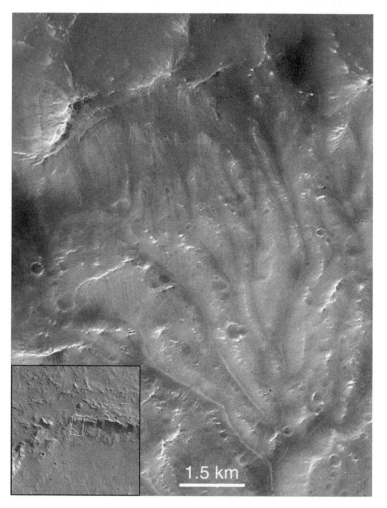

**Figure 8** System of water channels on the wall of Bakhuysen crater, reminiscent of drainage systems found on Earth

neither a fully satisfying biological nor a non-biological explanation for the observed clues. On balance, however, the proponents of life in the Martian meteorite(s) have the upper hand. What's more, the Mars life argument has grown even stronger following recent evidence for an ancient ocean on the fourth planet and the detection of methane in the Martian atmosphere.

# 4

# That Which Survives

It's hard not to be anthropocentric, or at least geocentric, when we think about the possibilities for life elsewhere. We know of only one kind of biology – that based, fundamentally, on carbon, with water as a solvent. So, when looking for life on other worlds, though recognising that exotic forms are possible, we first seek out the essentials that underpin all organisms here on Earth – organic (carbon-based) chemicals, water, and some suitable energy/nutrient source, or combination of sources, that might spark life into existence and later sustain its metabolism.

## Water, water, anywhere?

Of these three key ingredients, water has always topped the list in astrobiology, even before there was a science with such a name. When astronomers first pointed their telescopes at Mars, and at other worlds such as the Moon, they wondered if the dark patches they saw might be oceans. And if they convinced themselves that these regions were watery, they wondered what creatures might live on their shores or in their depths. When Percival Lowell thought he saw canals on Mars, he assumed there must be canal-builders and so the myth of intelligent Martians, which led to so many tales of extraterrestrial visitation and romance, was spawned.

Nothing much has changed in that regard. The search for alien life remains, first and foremost, a search for alien water – that most precious and unique of cosmic commodities. And, in particular, the astrobiological significance of Mars is predicated on the existence of sufficient water on the Red Planet, now or in the past.

Our perceptions about Mars have changed radically over the centuries, and our hopes for finding life there have swung wildly back and forth, even over the past few decades. The first spacecraft to fly past the Red Planet, in the early 1960s, led us to believe that it was a place as dead and desolate as the Moon. But these early probes – Mariners 4, 6, and 7 – gave a misleading impression because their cameras, by a twist of fate, happened to be trained on some of the most barren and uninteresting areas of Mars. Nor did their instruments have the sensitivity to pick up the detail which would have told a different story. Only with the arrival in orbit of Mariner 9 in 1971, and continuing with the Viking missions and later probes, has it become clear that Mars is a varied and still active world, rich in geological complexity. Recent observations have pointed to an abundance of water ice just below the surface and at the Martian poles, even some surface liquid water (see figure 9), and, tantalisingly, certain gases, notably methane, in the atmosphere which some process is continually replenishing.

It's true that, by human standards, Mars today is less than welcoming; if you were to land on its surface it would seem a disturbingly hostile place. The air is unbreathable, the surface is scoured by lethal solar radiation, and a pot of water would evaporate in minutes if it didn't freeze first, because both the atmospheric pressure and the average surface temperature are very low. At higher elevations, where the pressure drops even further, the water in the

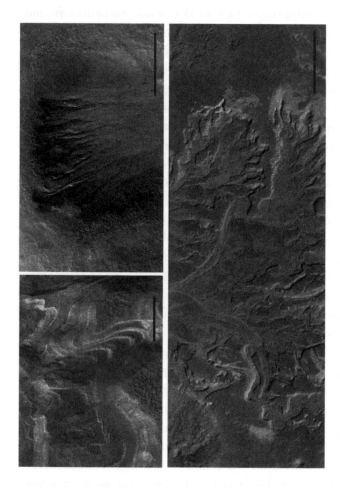

**Figure 9** Evidence of liquid water flowing and forming ponds on the surface of Mars. Top left: Layered sediments in Hellas Planitia, probably deposited on the floor of an ancient lake. Top right: Gullies in a crater wall. Bottom: Inverted relief of fossilised river channels forming a fan-like structure. Scale bars are 500 metres.

pot would bubble and sputter away, due to the greatly reduced boiling point of the liquid.

Although studies on the survival of Earth organisms under simulated Martian conditions point out how challenging the surface of Mars is to any kind of life we know, researchers such as Andrew Schuerger from the University of Florida and Charles Cockell from the British Antarctic Survey have demonstrated that Earth-like organisms could survive if covered by just a thin layer of soil.[27] In other work, Benjamin Diaz showed, in laboratory experiments designed to replicate Martian conditions, that potentially the most life-limiting factor on Mars is a paucity of surface or near-surface liquid water.[28] On the other hand, it's known that terrestrial microbes can get around this problem, as made clear by organisms that live and reproduce in the Antarctic Dry Valleys. Imre Friedmann's pioneering research on these primitive rock-dwellers suggests a way that Martian counterparts might be able to cope perfectly well with the challenges of their environment.[29]

## Mars in the past

The surface of Mars today seems quite hostile to life. But in the good old days, around 4 billion years ago, at about the time that ALH 84001 formed, the fourth planet was warmer, wetter, and probably endowed with a thickish carbon dioxide atmosphere. Alberto Fairén from the Universidad Autonoma in Madrid, Spain, and now at the NASA Ames Research Center, has identified ancient shorelines on Mars, and concluded that an ocean once sprawled across a third of the planet's surface.[30] Fairén's results were criticised at first because the margins of this supposed great body of water didn't seem to be continuous. However, in 2007, a study by a group of scientists from the University of California at Berkeley led by Taylor

Perron solved the mystery of the offset shores by proving that long ago the Martian poles had tipped fully 50 degrees from their original orientation.[31] This discovery not only explained why the old ocean margins were in places out of alignment; it also accounted for some remarkable erosional features on Mount Olympus, the biggest volcano on Mars, by tying them to climatic changes triggered by the axial shift.

Further evidence of a primordial ocean was published in 2009, based on data collected by the Mars Odyssey probe.[32] A gamma-ray spectrometer on the orbiting spacecraft can measure the concentrations of various chemical elements as much as 30 centimetres under the Martian surface. Elevated concentrations of potassium, iron, and thorium below the shoreline, and lower concentrations above it, suggest that these elements were leached out of the soil by run-off water and became enriched in muddy sediment on the floor of a standing ocean.

Mars once had a substantial atmosphere, comparable in density to our own. But the planet's gravitational pull (with a surface strength only about a third that of Earth's) was too feeble to hold on to this massive gassy blanket and much of it gradually seeped away into space. As Mars lost the bulk of its ancient atmosphere, it also lost most of its greenhouse heating capacity, and so grew increasingly frigid and dry.

Persistent cold, arid conditions became the Martian norm. Yet, occasionally, they were interrupted by periods of spectacular activity, lasting perhaps tens of thousands of years. During these hiatuses, catastrophic flooding visited the northern plains, forming bodies of water that ranged in size from small lakes to oceans 20 times the size of the Mediterranean. It still isn't clear what caused these deluges, though various theories have been touted. James Dohm of the University of Arizona at Tucson, for example, is among those who've argued that they coincided with episodes of volcanic activity.

Volcanoes could have quickly pumped impressive quantities of greenhouse gases into the atmosphere, leading to global warming and associated melting of the Martian ice. The great size of Mount Olympus, with a base that would blanket Arizona, and that of its sister volcanoes in the Tharsis and Elysium Planitia regions, support this notion. Another idea is that periodic variations in the axial tilt of Mars, or of its orbital parameters, might have led to sudden melting of large areas of polar ice.[33] This melting would have released large amounts of water vapour, triggering a temporary greenhouse effect and lifting temperatures across the planet. It's possible that the two processes were coupled and happened at the same time.

## Survival strategies

In general, however, over the aeons, Mars has become increasingly frigid. Planet-wide falling temperatures suggest that, for the most part, any remaining liquid water has retreated further and further underground. Anything resembling Earth organisms, dependent on this water, must also have moved deeper, under the permafrost, perhaps continuing to thrive, as some terrestrial microbes do, in granitic and basaltic groundwater thousands of metres below the surface. Another possibility is that some Martian life found a way to adapt to the extreme cold and desiccation of the topsoil, for example using a kind of intracellular antifreeze as described earlier.

In a 2005 paper, a research team elaborated on various scenarios for the evolution and persistence of life on Mars.[34] It suggested that as the environmental histories of Earth and Mars increasingly diverged after their first few hundred million years, so too might their biological trajectories. Any microbes that still survive on Mars near the surface must have become extreme psychrophiles ("cold-loving" organisms), either tapping inorganic food sources in otherwise nutrient-poor surroundings, or using photosynthesis in very specific habitats, such as the fringes of the polar ice caps. This same investigation looked at the possibility of Martian life that has evolved to cycle between periods of activity and of long dormancy, in which case microbes could be present in mostly dormant forms near the surface and in active forms in more clement, protected environments at lower levels. The periodic availability of liquid water on the surface might have fuelled bursts of biological activity (much as some desert regions on Earth bloom exuberantly in the days immediately following a rare downpour), as well as giving the opportunity for evolutionary progress driven by natural selection.

It's entirely possible then that Mars experiences a reawakening of part of its biosphere during spells when a surge of liquid surface water becomes available, and a hunkering-down of life when lengthier cold and dry periods prevail (as is the case on Mars right now). Such a situation isn't unlike what we see on Earth, when some organisms hibernate or become torpid during the cold days of winter, and reanimate with the coming of spring. On Mars the intervals between warm spells may be a tad longer – millions of years instead of a few months. But, at least as far as bacterial spores are concerned, there seems to be no practical limit on how long they can remain dormant. Raúl Cano and Monica Borucki reported in the journal *Science* on the isolation of a viable strain of *Bacillus sphaericus* from an extinct bee trapped in 25- to 30-million-year-old amber,[35] while controversially, a paper by Russell Vreeland and colleagues in *Nature* described the finding of a viable 250-million-year-old bacterium inside a salt crystal in New Mexico.[36] If organisms on Earth can survive as spores for tens or even hundreds of millions of years without the encouragement of any special evolutionary pressures, then there seems no reason why Martian organisms couldn't have developed similar or greater capacities for suspended animation.

## Liquid refreshment

How times have changed. In 1964, Mariner 4 seemed to speak of a world as dry as a lunar dustbowl. Now, four decades on, compelling new evidence has come to light of the presence of recent, and even contemporary, running water on Mars. In 2000, Michael Malin and Kenneth Edgett of Malin Space Science Systems used imagery from the Mars Global Surveyor spacecraft to identify more than 120 locations where it appears that water has seeped into freshly-cut gullies and gaps on the Martian surface.[37] Their announcement prompted a NASA press conference at which the authors expressed surprise at uncovering so many wet locations, a high percentage of which were on steep slopes in the Martian southern hemisphere that receive little sunlight. The interpretation was clear: there must be liquid groundwater relatively close to the surface, which occasionally breaks through some kind of frozen ice barrier, triggering a mudslide that rushes downhill. Malin and Edgett suggested that the features they identified could have been due to water releases anywhere from a few million years ago to yesterday.

Six years later the same researchers compared up-to-the-minute surface images with older photos of Mars obtained by the same spacecraft. What they found was astonishing. The new pictures revealed bright new sediment deposits in one of the gullies, showing beyond doubt that material had been swept down the slope at some point *in the past ten years*.[38] The shape of the deposits indicated that flowing water, bursting through the canyon wall, was the culprit. And the length of the deposits – a few hundred metres – suggested that the water remained in a liquid state for at least several seconds until, presumably, it evaporated into the thin Martian air (see figure 10).

This remarkable discovery brings fresh urgency to a number of key questions. How much liquid water lies below ground level on

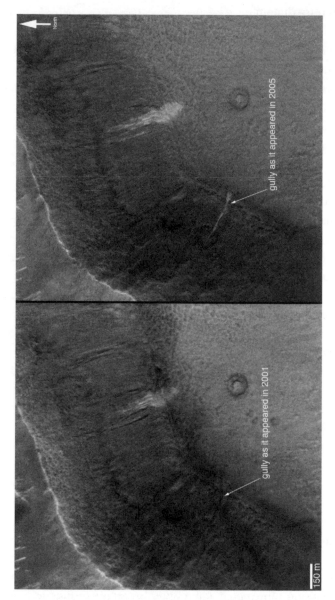

**Figure 10** New gully deposit in a crater in Terra Sirenum, Mars. The gully is thought to have recently formed by some type of process involving liquid water.

Mars? How close is it to the surface? And does it, in fact, provide a habitat for life?

We know now that water is present on Mars and in much greater abundance than could have been imagined after the early flyby missions. Water in association with heat (for example, in the vicinity of hot springs and undersea vents) provides a nurturing environment for life on Earth. Evidently, it is available to do the same on Mars, a conclusion bolstered by the fact that water doesn't need much heat to remain in liquid form, especially if it's rich in salts.

Water changes the face of Mars, even from year to year. But the Red Planet is restless in other ways that lend credence to the argument for life.

## Mysterious emanations

In 2004, Mars stunned the scientific world again. Three different research groups reported the same finding: the detection of methane in the Martian atmosphere.[39] This was totally unexpected and immediately raised the possibility that the gas had a biological source on or beneath the surface.

On Earth, most atmospheric methane comes from the metabolic activity of microbes known as methanogens. The key point is that on Mars (as on our own planet), atmospheric methane is extremely unstable because it's continually being broken up by ultraviolet rays from the Sun and chemical reactions with other gases. The average lifetime of a methane molecule in the Martian atmosphere is no more than 400 years,[40] which means the gas must be continuously replenished if it isn't rapidly to disappear. Something is producing methane on Mars today; the big question is, exactly what?

Although the measured methane concentrations are low, they're high enough to be unmistakably above background levels. Any doubts the detection was real were dispelled by the fact that it was made by a trio of independent and experienced research groups. These were led by Vladimir Krasnopolsky from the Catholic University of America, Michael Mumma from the NASA Goddard Space Flight Center, and Vittorio Formisano from the Institute of Physics and Interplanetary Science in Rome, Italy. The results provided by Mumma and Krasnopolsky came from ground-based observations, while Formisano's group used data collected by the European Mars Express probe. The spectrometer onboard Mars Express was also able to measure longitudinal (east-west) variations in the methane concentration, which showed peaks over certain areas of the planet. These peaks suggested localised sources of the gas, perhaps associated with volcanic activity. Since volcanic activity produces little methane by itself, the serious possibility arose that the emanations were due to life in and around hydrothermal vents.

Astrobiologists around the globe quickly grasped the significance of this new discovery and began to theorise on the various means, both biological and chemical, by which the methane might be produced. One intriguing possibility was through the recycling of deeply buried kerogen, an insoluble organic material, by a long heating process known as thermogenesis. This idea, suggested by Dorothy Oehler of the Johnson Space Center, implies a delayed type of life-generated methane formation.[41] The heat needed for the thermal processing could come from molten rock underground or meteorite impacts. Although this heating might also encourage water-rock interactions in the subsurface that would release methane even in the absence of biology, such contributions would be quite small.

## Microbial methane

In 2006, Tullis Onstott from Princeton University and his colleagues looked at a number of other ways the Martian methane might come about.[42] They considered a variety of non-biological processes, such as those taking place near underground heat sources. They also looked at methane-containing hydrates (water-rich ice structures that trap methane) and the microbial generation of methane above the planet's permafrost layer.

The Onstott group focused especially on comparing the situation on Mars to environmental conditions it had investigated in the deep-lying rocks of the Witwatersrand Basin in South Africa. In these rocks, microbial-produced methane was prevalent, a fact confirmed by DNA sequencing. The major nutrient source for the methane-producing microbes was hydrogen gas, produced either by water-rock interactions or the radioactive splitting of water in the crust. The existence of this kind of ecosystem had already been demonstrated in deep basalt rock at the Hanford nuclear site in Washington State, USA.[43]

What Onstott and his colleagues showed was that hydrogen produced by the underground radioactive breakdown of water could explain the methane concentrations seen on Mars *if* the hydrogen were microbially converted to methane. Or the methane could derive from near-surface, hydrate-rich ice at low latitudes, where most of the gas seems to originate. In this case the methane could either be from a chemical or biological source. Steve Clifford from the Lunar and Planetary Institute previously pointed out the likely presence of large amounts of carbon compounds trapped in water-rich ice structures on Mars. On Earth, the methane trapped in these structures is overwhelmingly derived from biological processes. The Onstott group urged that future missions to the Red Planet should be designed to measure abundances of methane, helium, and hydrogen, and to determine the carbon and hydrogen isotopic composition of methane and higher hydrocarbons, such as ethane, using an instrument that could be attached to a rover. One of the most intriguing results of their research was that the methane, if microbial, would have to come from *above* the permafrost, requiring organisms to be located relatively close to the surface.

Other organisms might be present that oxidise methane, either using oxygen, or in reactions in which sulphate compounds are converted to sulphides. Either process would cause changes in the ratio of the two main isotopes of carbon (carbon-12 and carbon-13) in the residual methane, which we should be able to detect in the future. There is potentially an enormously large biosphere present a few metres below the surface of Mars – a biosphere which the Viking mission may have not been able to access since it was only scratching the surface of the uppermost soil layer.

Michael Mumma, one of the original discoverers of methane on Mars, was again in the news in October 2008 when he announced at an American Astronomical Society meeting that the region known as Nili Fossae was among the methane-emitting hotspots, with peak concentrations of up to 60 parts per billion. This is a big deal, because an average global level of ten parts per billion and a lifetime of hundreds of years would mean that a few hundred tonnes of methane are entering the atmosphere each year – equivalent to the amount of methane that a few thousand cows on Earth release.[44]

The position regarding the detection of methane parallels that of the Viking and ALH 84001 controversies. The most straightforward explanation is life, and if Earth were involved, instead of Mars, there would be no doubt which interpretation scientists would choose. What complicates the issue is that we're dealing with a foreign planet whose history and processes we still far from fully understand.

## Formaldehyde

Another big surprise came at the Mars Express Science Conference at Noordwijk in the Netherlands in February 2005. Vittorio Formisano, principal investigator of the Planetary Fourier Spectrometer onboard the Mars Express spacecraft, announced details of the detection of formaldehyde (chemical formula HCOH) in the Martian atmosphere. The mere presence of formaldehyde wasn't surprising, since this chemical is a natural oxidation product of methane, and its discovery was expected on the heels of the methane disclosure. What stunned the scientific world was the sheer quantity of formaldehyde involved: 10 to 20 times greater than that of the methane.

Formaldehyde is much more unstable in the Martian atmosphere than methane; it survives, on average, only 7.5 hours before breaking apart. This means that if the detection was correct and formaldehyde was being formed from oxidised methane, Mars must be producing roughly 2.5 million tonnes of methane per year. This amount would be very difficult to explain in terms of inorganic processes and would strongly favour a microbial explanation.

The catch is that the detection of formaldehyde is nowhere near as secure as that of methane. The measurements reported to date are right at the limit of the instrument's capability and haven't been confirmed by any independent research groups. If the finding is eventually verified, it would suggest a flourishing microbial biota, large enough in population and spatial extent to produce the required massive methane degassing. This conclusion, however, would have to be squared with other considerations, such as the apparently nutrient-poor Martian environment and the lack of organic molecules found by the Viking GCMS, which argue for a more restricted biosphere.

Interestingly, the formation of methane can also run the other way: some microorganisms on Earth break down nitrotoluene derivates, producing formaldehyde, which can then generate methane as a degradation product. This interpretation is more consistent when examining the relative amounts described of both gases.[45] In any event, the number and variety of possible biogenic explanations for the presence of methane and formaldehyde in the atmosphere of Mars, and the weaknesses of the rival abiogenic hypotheses, very clearly support the case for life.

# 5

# Phoenix and the Future of Mars Exploration

Greek mythology tells of a wonderful bird called the phoenix, which lived for many centuries before bursting into flames at the end of its life. From its ashes a new bird arose.

NASA's Phoenix mission was born from the ashes, so to speak, of two spacecraft designed to explore the surface of Mars, but which never made it. The Mars Polar Lander (MPL) crashed somewhere in the south polar region of the planet in December 1999, possibly due to a software error during descent, while the Mars Surveyor 2001 Lander was cancelled by NASA in the wake of the MPL failure and that of the Mars Climate Orbiter. Phoenix, the first of NASA's smaller, lower-cost "Scout" missions designed to complement its bigger Mars projects, used duplicates of instruments from MPL, and three instruments from the Mars Surveyor 2001 Lander which had been kept in storage. Phoenix's target was a subpolar plain, not far away from the frozen water of the north polar ice cap (see figure 11).

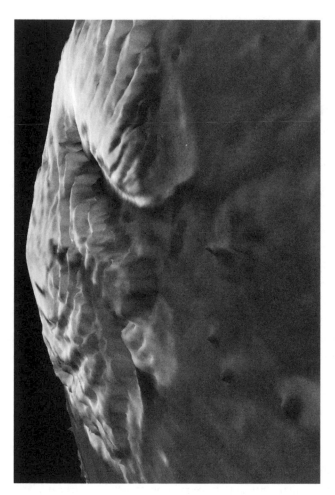

**Figure 11** Three-dimensional view of the North Pole Ice Cap on Mars taken by the spacecraft Mars Global Suveyor. The ice is mostly water ice as confirmed by remote sensing. The cap is roughly 1,200 km across and as much as 3 km thick. With an average thickness of 1 km, the volume of ice is about 1.2 million cubic kilometres, roughly half that of the Greenland Ice Cap on Earth

## Aims and devices

Launched on 4 August 2007, Phoenix touched down safely on a low-lying, subpolar expanse of Mars known as Vastitas Borealis ("northern waste") on 25 May 2008 (see figure 12). Its goal was to gather geological and meteorological data with a view to assessing whether this region is, or ever had been, a suitable dwelling place for microbes. Although not actually a life-detection mission, Phoenix was designed to search for organic compounds and other evidence to support or discredit the notion of past or present habitation. It was equipped with a robotic arm to dig through the protective layer of topsoil to any water ice below and bring both soil and water ice to the lander platform for analysis.

The two most important instrument packages aboard Phoenix were TEGA (Thermal and Evolved Gases Analyzer) and MECA (Microscopy, Electrochemistry, and Conductivity Analyzer). TEGA contained eight miniature ovens and a mass spectrometer. The temperature of the little furnaces could be raised to $1000^\circ C$ (1800°F) to vaporise any ice and other volatile materials in the sample into a stream of gases. These gases were then fed to the mass spectrometer, to have the masses and concentrations of their molecules and atoms identified. With a sensitivity of ten parts per billion, the mass spectrometer was designed to sniff out even tiny quantities of organic chemicals that might exist in the ice and soil.

MECA was essentially a wet chemistry lab, transplanted from the Mars Surveyor 2001 Lander and charged with the task of character-ising the Martian soil, much as a gardener might test the soil in her garden. By dissolving small amounts of soil in water, MECA could determine the pH (acidity and alkalinity level), and the abundance of minerals containing magnesium, sodium, chloride, bromide, and sulphate, as well as dissolved oxygen and carbon dioxide. MECA was also equipped with little probes to measure the soil's water and

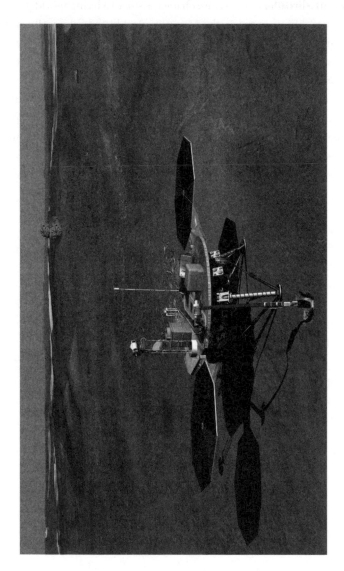

**Figure 12** Artist's rendition of the Phoenix lander on the arctic plains of Mars with the polar water ice cap in the far distance

ice content, and the ability of heat and water vapour to travel though it. Four single-use beakers in MECA could each accept one small sample of soil. The spacecraft's robot arm started each experiment by delivering a sample into a beaker, which then began testing for carbonates, sulphates, and oxidants.

The MECA package also included two microscopes. An optical microscope enabled soil grains to be examined to help determine their origin and mineralogy. An atomic force microscope, able to resolve detail as small as one ten-billionth of a metre across, was designed to detect water-containing substances, such as clay minerals, which might indicate past liquid water in the Martian Arctic. The optical microscope, although having a lower resolution of four millionths of a metre, could illuminate samples with red, green, blue, and ultraviolet light in different combinations to reveal subtleties in the soil and water-ice structure and texture. In theory, the microscopes could see microbes, which on Earth have an average diameter of about one millionth of a metre. However, the optical microscope would only detect them if, by sheer luck, a bacterial colony was imaged directly, and even then it would be questionable whether a decisive call could be made.

## News from the northern plains

Early results from Phoenix were intriguing. One of the trenches dug by the robot arm was nicknamed "Snow White" because of the coating on the exposed soil. Within a few days the white material disappeared, showing clearly that it had been water ice (the only substance that would behave in this way). This detection of near-surface ice is exactly what had been hoped for, and the very reason the landing site had been chosen.

Another Phoenix discovery proved even more exciting in the quest for habitability. Chemical analysis of the soil turned up a pH value between 7 and 9, meaning that it was slightly alkaline. Also detected were the elements magnesium, sodium, potassium, and chlorine, which on Earth serve as important mineral nutrients. Some previous studies had concluded that the Martian surface might be too acidic or salty to provide favourable conditions for life. But this new analysis showed that soil in the subpolar region has the same basic chemistry as garden soil. The soil, excavated from the top few centimetres of a patch of ground fittingly called "Wonderland," was a close match to what's found in the Antarctic Dry Valleys on Earth, where a small but resilient microbial community makes a living. With a little fertiliser thrown in, vegetables like asparagus and turnips could probably be grown in it – an interesting prospect for any future green-fingered colonists who'd like to try their luck at horticulture on Mars.

Phoenix co-investigator Samuel Kounaves of Tufts University was quick to make the point that soil conditions at the spacecraft's landing site looked distinctly bio-friendly. And, with hindsight, that isn't so surprising. Mars, like Earth, surely has many different microenvironments. There's no reason to suppose that the sulphur-rich and likely acidic and salty conditions found at some locations near the Martian equator span the entire planet, any more than the Badlands in South Dakota or the Gobi Desert are representative of our own world.

But Mars, it seems, loves to play games with us. No sooner do we uncover evidence that favours life, than something else crops up to make us think again.

## Good news, bad news

A few days after Phoenix had sparked talk of soil good enough to support asparagus, its MECA instrument suggested that there might be perchlorates in it,[46] which was soon confirmed by the TEGA package.[47] Perchlorates are highly oxidising compounds which, at first glance, don't bode well for the prospects of native biology.

However, Steven Benner, a chemist at the Foundation for Applied Molecular Evolution in Gainesville, Florida, suggested otherwise. Two key concepts in physical chemistry are kinetics and energetics. Kinetics describes how quickly a chemical reaction takes place; energetics tells how much driving force is behind the reaction. Benner likens these concepts to two cars on top of a hill, one with good brakes, the other with bad. Both have the same potential energy and will eventually reach the bottom of the hill, but the second will get there much more quickly. Perchlorates are like the car with good brakes; although they're very energetic, they're also highly stable – so stable that any organic molecules exposed to them can survive for a very long time. In fact, a powerful but stable oxidant is beneficial to life; there are even some microbes on Earth that tap perchlorates as a food source, using enzymes to increase the rate at which the intracellular perchlorate reactions take place.

At the time of writing, there's no definite evidence that Phoenix found any organic compounds, and the perchlorates may be the reason why. While perchlorates are stable under Martian environmental conditions, they're highly combustible when heated in an oven such as those in the TEGA instrument. Any organics would be oxidised to carbon dioxide and thus rendered undetectable. The same reaction might also have affected the Viking Gas Chromatograph Mass Spectrometer (GCMS) experiments.

Some perchlorates have interesting properties that would be potentially useful for organisms in the polar regions of Mars. Magnesium perchlorate, for example, is very hygroscopic (attracting liquid water directly from the atmosphere), stable at low temperatures, and, in saturated solution, able to remain liquid down to temperatures of about -70°C. This is above the mean Martian temperature, and so allows for the possibility of liquid briny water at or

near the surface. One of the struts of the Phoenix lander had a number of blobs on it, which were interpreted to be liquid water
splashed on to the metal during the last few seconds of landing.[48]
This salt-rich water may even be involved in the fresh-cut gullies on
Martian hill slopes observed by Malin and Edgett.

The acid equivalent of perchlorates – perchloric acid – is also
strongly hygroscopic. Its properties are similar to those of hydrogen
peroxide, although the melting point of a perchloric acid-water
mixture is not quite as low. Such a mixture might be able to take
on some of the biological functions suggested by the hydrogen
peroxide-water hypothesis discussed earlier.

Phoenix found clay minerals and calcium carbonate at its landing
site, adding further to the body of evidence that Mars had a watery
past. Calcium carbonate tested positive in both the MECA and
TEGA instruments, and carbonate rocks have also recently been
seen from orbit in another region of Mars, Nili Fossae.[49] The presence of this chemical (which is found on Earth in various forms
including limestone and chalk) is very significant because it confirms
the neutral or slightly alkaline pH of large swathes of Martian soil. It
also suggests that a lot of the water on Mars was, and may still be,
similar to that in Earth's oceans. Acidic water, on the other hand,
would have dissolved all the carbonates.

Phoenix was the first spacecraft to detect snow falling in the
Martian atmosphere.[50] Even more surprisingly, a diurnal cycle was
recorded at the landing site. Ice is deposited, mostly as frost, at night
and sublimates in the morning hours. Near midnight, ice clouds
form and precipitate a portion of the atmospheric water back to
the surface. Water is also plentiful in the soil. A shallow water ice
table was uncovered by the robotic arm of the Phoenix lander
at a depth of 5 to 18 centimetres (less than 1 foot), beneath which the
soil pores are completely filled with water ice. The characteristic
clumping of the soil at the landing site indicates that some of that

water was liquid, most likely in the form of a briny film on the surface of minerals. Such films of water may be all that microorganisms need to make a living on the Red Planet.

Mars, through the eyes of Phoenix, is a diverse world with a subpolar region that was once, and may still be, habitable. Other research supports that notion and suggests that this region was at various times covered by shallow oceans.[51]

Eventually, after about six months of operation the Martian winter took its toll on the spacecraft's solar-powered batteries and the lander fell dead in November 2008.

## Next steps

Phoenix couldn't, and was never intended to, prove the case for life on Mars. However, two future missions may do exactly that. NASA's Mars Science Laboratory (MSL), a sophisticated rover scheduled for launch in 2011, will be the next best thing to having a human geologist on Mars backed by a chemical laboratory. Its goals are to look for clues about whether life ever arose on the planet, to add to our knowledge of Martian climate and geology, and to pave the way for human exploration. MSL is the size of a car – much bigger than the Mars Exploration Rovers "Spirit" and "Opportunity" – and carries equipment to probe for organic material and chemical signatures of biology. It also has a thermal-nuclear battery, powerful enough to let MSL roam many kilometres from its landing site over a period of at least two Earth years.

At the heart of Mars Science Laboratory's scientific payload is an instrument suite called Sample Analysis at Mars (SAM), built to analyse organic chemicals and gases from both atmosphere and soil. SAM will be able to pin down the carbon isotope ratios of carbon dioxide and methane, which will help determine whether the

**Figure 13** NASA's Mars Science Laboratory is a mobile robot for investigating the past or present ability of Mars to sustain microbial life. This picture is an artist's concept, portraying what the advanced rover would look like in Martian terrain. The MSL rover is much bigger than the previous rovers, with the mast rising to about 2.1 metres above ground level

methane plumes on Mars are of geochemical or biological origin. It should also shed some light on the hydrogen peroxide–water hypothesis of Martian life, as it will be able to detect hydrogen peroxide and any other strong oxidisers. Although not designed to look for life directly, MSL is the first mission that is truly superior in many analytical and experimental capabilities to Viking (see figure 13).

Four locations have been short-listed as possible MSL landing sites. Eberswalde Crater, rich in clay minerals, would allow scientists vicariously to sift through the sediments of a delta, where a river once flowed into a lake, for signs of life-related carbon chemistry. Gale Crater would challenge MSL with a drive up a five-kilometre sequence of layers to study the transition from environments that produced clay minerals to later environments that left behind sulphate deposits. Thirdly, there is Holden Crater, where running water once carved gullies deep into the sediment, and both catastrophic flood deposits and lake deposits are exposed. Finally, Mawrth Vallis, a flood channel near the Martian highlands, is intriguing because it holds different types of layered clays, revealing the shift from a wet to a dry environment. A fifth potential landing site, in the Nili Fossae region, has also come into the reckoning with the recent discovery that it is a source of methane plumes.[52] MSL would be able to sniff the methane to determine whether its carbon isotope ratio seems biological in origin or not, and gauge the amount of methane being released.

Even more promising in the quest for Martian life is the European Space Agency's ExoMars mission, scheduled for launch in 2018, and consisting of an orbiter, a descent module, and a rover. The orbiter will act as a data relay satellite, while the rover will be able to trundle several kilometres over the planet's surface. New developments in technology will allow the rover to operate autonomously using onboard software and optical sensors. However, the most exciting aspect of the mission is the 40-kilogram exobiology payload, complete with a lightweight drilling system that can bore up to two metres below the surface. A key part of the payload is the Urey Instrument being developed by Jeffrey Bada, of the University of California at San Diego, and his team. Urey contains the Mars Organic Detector, which is fine-tuned for analysing the building blocks of life as we know them: amino acids, nucleotides and nucleic acids (such as the DNA and RNA in our genetic code), and amino sugars. It can also detect polycyclic aromatic hydrocarbons, which were previously found in the Martian meteorite ALH 84001.

A second component of Urey is the Mars Oxidant Instrument, built to determine the oxidation state of the Martian soil and look for specific oxidising compounds. Hopefully, it will clear up once and for all the riddle of the extreme soil activity found by Viking.

In 2013, NASA plans to launch the second spacecraft in its Mars Scout Program. Known as MAVEN (Mars Atmosphere and Volatile Evolution), this mission, led by Bruce Jakosky of the University of Colorado, will look further at the climate and habitability of Mars, especially in regard to dynamic processes in the upper Martian atmosphere.

Scientists in the United Kingdom have proposed a European astrobiology-focused mission called Vanguard, which would build on technology developed for the ill-fated Beagle 2 probe, lost following its separation from Mars Express in 2003. Vanguard would comprise a lander, a micro-rover similar to the successful Sojourner carried by Mars Pathfinder, and three ground-penetrating moles mounted on the rover, which could each burrow five metres below the surface to search for any life that might be sheltering at such depths.

In the shorter term, Russia plans to launch a spacecraft to the larger of the two Martian moons, Phobos, in late 2009. Scheduled to land on Phobos in 2010, it will collect soil samples and return them to Earth two years later. The mission preparations are on a tight schedule and if the launch date had to be delayed, another launch window would open in 2011. Also, a spacecraft that could grab a piece of Mars itself and bring it home for analysis has been under discussion for years. The current NASA timetable calls for a sample-return attempt no sooner than 2020 – the timing uncertain because of the mission's high price tag and the agency's other budget commitments.

## Toward a final proof

Within one or two decades, thanks to robot probes now in development, we may know for sure if Mars is, or ever was, inhabited. Already the evidence, though circumstantial, is compelling. For several hundred million years, in its youth, Mars was warmer and wetter. Temperatures must have been typically above freezing and liquid water pooled into great lakes and oceans. A planetary magnetic field shielded the surface from harmful solar radiation, and volcanic activity provided numerous local sources of energy and heat. Early in the history of the Solar System, Mars was likely more suitable than Earth as an incubator of life because our own world took much longer to cool down from its hot and fiery beginnings. Earth also experienced a cataclysmic impact with a Mars-sized object shortly after its formation. That collision hurled part of Earth's mantle (the partially molten rocks between the core and the crust of a planet) into space, providing the raw material from which the Moon would later coalesce.

Meteorite exchange was common when the Solar System was young, offering an alternative route for microbial seeding of Mars. Given that both the third and fourth planets were life-friendly some four billion years ago, and that it appears microbes can survive long space voyages if protected by even a thin layer of rock, it is difficult to imagine that the infant Mars could have failed to acquire life by one means or another.

The testimony of Martian meteorites, and especially the magnetite chains in ALH 84001, strongly favours the past presence of bacteria on the Red Planet. Chains of magnetotactic bacteria would be a logical adaptation to the influence of a global magnetic field, just as they were on Earth. The detection of methane in the Martian atmosphere supports the case for biology. Most likely, the methane comes either from clathrates (substances with cage-like molecules in which methane could be trapped) formed by life processes billions of years ago or from present-day microbes associated with volcanic heat sources.

Astrobiologists have become comfortable with the idea that primitive life, at least, is common throughout the universe. In the main this is because it sprang up so quickly and easily on Earth, helped along by a few basic ingredients which may occur routinely elsewhere. Given that Mars once had those same ingredients there's every reason to suppose it had a thriving biosphere. Gradually, conditions grew tougher. Yet we know how incredibly resourceful, adaptive, and hardy terrestrial microbes can be. Take, for example, the hyper-dry Atacama Desert in Chile – one of the best natural analogues to a Martian environment on Earth. Despite having a mean annual precipitation of less than two millimetres, it still manages to host communities of photosynthesising bacteria in its rocks.[45] Some of these rock-dwelling organisms, called "endoliths", survive by using the salts on the surface of their cells to extract tiny amounts of water vapour from the atmosphere (see figure 14).

Other organisms on Earth draw nourishment from the iron and sulphur in acidic mine drainage, which would be deadly to higher animals like humans. Iron and sulphur are common elements on the Martian surface, and native creatures there would likely become

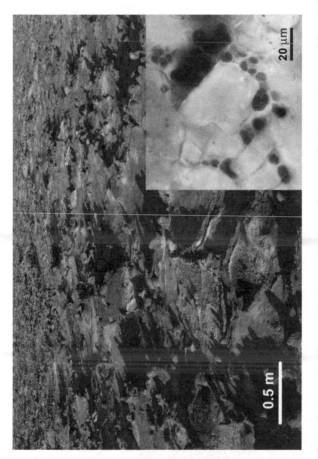

**Figure 14** Top: Halite (sodium chloride – a table salt) crusts in the very dry core of the Atacama desert in Chile, which are colonised by cyanobacteria (inset) that live within the rocks and take advantage of the hygroscopic properties of the mineral to obtain liquid water from the atmosphere. Chloride-bearing deposits on Mars which could have similar properties to the salt crusts in the Atacama desert could provide a habitable niche for well-adapted microorganisms

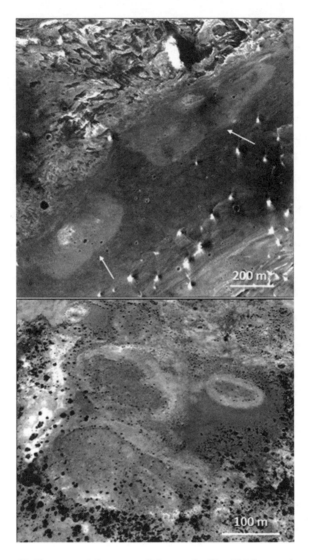

**Figure 15** Upper panel shows two light-toned, ellipsoidal features (arrows) on Mars that Allen and Oehler interpret as remnants of ancient spring deposits. Lower panel shows extinct springs from the Dalhousie Complex in Australia, characterised by their light-toned, ellipsoidal structure[53]

adapted to use them in their metabolism. A variety of microbial communities spring up around terrestrial hydrothermal vents, and the same could be true for Mars. Carl Allen and Dorothy Oehler from the Johnson Space Center identified tonal anomalies in Arabia Terra on Mars (shown in figure 15), which they think might represent the mineral residue from ancient hot springs on Mars.

The bottom line is that if life ever got a foothold on Mars it probably still exists today. No matter that Mars is currently cold and dry; there is plenty of evidence that the planet goes through intermittent warmer, wetter periods, probably associated with volcanic outbursts, when long-dormant microbes could reanimate, multiply, and even evolve before the next big freeze sets in.

On Earth liquid water is plentiful, on Mars it is not. Under such circumstances, hydrogen peroxide mixed with water would be a perfect substitute for pure $H_2O$, because it would allow Martian organisms to survive the cold by keeping their cellular contents liquid and obtaining water directly from the atmosphere. Further, the same chemical that would protect a cell from overly active hydrogen peroxide would also shield the cell from harsh radiation, allowing microbes to survive on or near the surface. The proposers of the hydrogen peroxide–water hypothesis for life on Mars also suggested a photosynthetic pathway that would convert carbon dioxide, water, and light to methane and hydrogen peroxide.[54] Thus, the detected methane in the Martian atmosphere might be produced by microbes today that are perfectly adapted to the near-surface environment.

One way to support the case for such organisms on Mars might be with laboratory experiments. The results of these experiments could then be compared with observations made by Phoenix, Mars Science Laboratory, and other missions that employ a lander or rover on the Martian surface. Recently, a research team made up of Houtkooper and scientists from Washington State University and

NASA Ames has designed a set of laboratory tests to reveal how a chemical response could be distinguished from a biological one, which then could be compared to the measurements taken by the TEGA instrument package on board Phoenix or the SAM instrument on board Mars Science Laboratory.

Unfortunately, the Phoenix mission did not obtain the quality of data needed to provide an answer to the possible existence of such organisms, but perhaps the Mars Science Laboratory will be more successful. In a 2008 paper, the group concluded that the following observations would provide evidence of hydrogen peroxide-water life on Mars:

1. The detection of a chemical stabiliser of hydrogen peroxide or its fragments;
2. The production of excess heat and the detection of gaseous decomposition products of organic molecules, such as carbon dioxide, water, oxygen, and nitrogen (the decomposition product of hydrogen peroxide would only be water and oxygen);
3. A measured distinct change toward lighter isotope ratios in carbon and oxygen at temperatures at which organic molecules decompose.[55]

This last indication would be critical, because life-forms on Earth, and presumably elsewhere, prefer lighter atoms for the simple reason that they take less energy to metabolise.

Another way to confirm life on Mars would be to investigate methane emissions directly. As mentioned, Mars Science Laboratory mission scientists are now seriously considering Nili Fossae, where high methane emissions have been detected, as a landing site. If MSL or any other rover were to directly measure the isotopic composition of the methane, plus any other organic compounds that may be present, we would likely be able to distinguish between a biogenic and a chemical source, again because life favours the lighter isotopes.

Finally, the Urey Instrument of the ExoMars mission may prove the presence of life. If it were to detect large organic molecules typical of terrestrial organisms, such as chlorophyll, ATP, or RNA, and assuming that the possibility of contamination could be excluded, then this would clinch the case for Martian biology. But even if life on Mars didn't use many of the same biochemicals as on Earth, we might catch a glimpse of it. The reason is that Urey is designed to detect amino acids, which are the building blocks of proteins. If amino acids are detected, they will be further analysed by Urey's Micro-Capillary Electrophoresis Unit to determine their handedness. This could prove decisive, because life on Earth uses molecules of a single handedness, either left or right, not both at the same time. A clear dominance of left-handed amino acids would suggest a strong similarity between, and perhaps even a common origin for, life on Mars and on Earth (where all amino acids used in proteins are left-handed). Alternatively, if the detected amino acids were right-handed, this would be evidence of Martian life with an independent origin – potentially an even more thrilling discovery.

The endgame in our long quest for life on Mars has begun, and a positive outcome seems likely. But Martians, if they exist, may not be the only extraterrestrials in our cosmic backyard.

# Part II

**A**strobiologists have become much more broad-minded. There's a growing realisation that even on Earth plenty of life exists that would have seemed truly alien just a few decades ago. Other worlds around the Sun that, at first sight, look like non-starters from a biological standpoint are turning out to have regions on, below, or above their surfaces where we can easily conceive of some kind of organisms making a living. What's more, some of these worlds, ranging from hot Venus to frigid Titan, are presenting us with data that fit well with the extrapolation that they are inhabited. For much of the Space Age, Mars appeared to be our only realistic hope for finding extraterrestrials within striking distance. Now astrobiologists have at least half a dozen planets and moons within the Solar System earmarked for exploration and life-oriented searches. Meanwhile, a whole universe of uncounted worlds awaits our investigation.

# 6

# The Clouds of Planet Hell

**N**o planet in the Solar System is more hostile to life as we know it than Venus. But it wasn't always regarded that way. Less than a century ago, the Swedish chemist Svante Arrhenius spoke of a world "dripping wet" and "covered with swamps," much as Earth must have been in the great coal-forming period of the Carboniferous, some 300 million years ago. Perpetually veiled as it is by thick white clouds, it's easy to see why some people thought that Venus might resemble a lush, steamy jungle.

Yet that all-concealing white blanket turned out to be neither unbroken rain cloud nor dense watery mist. Spectroscopic observations made from Earth, beginning in the 1920s, and then close-up studies by spacecraft, from the mid-1960s on, revealed that the vast bulk – more than 96% – of Venus' atmosphere consists of carbon dioxide. What's more, the Venusian atmosphere is 90 times thicker than Earth's, so that standing on the surface of the second planet you'd be subjected to a crushing pressure similar to that several hundred metres underwater. Unbearable heat, too, is a permanent feature at ground level. The massive amount of carbon dioxide in the atmosphere has led to a runaway greenhouse effect that makes Venus hotter even than the nearest planet to the Sun, Mercury. Day and night, over the entire surface, the temperature on Venus varies little from a blistering 460°C.

## Climate catastrophe

Something went horribly wrong with Earth's so-called sister planet. Venus is similar to our world in size, mass, and rocky composition. And although it's a little under three-quarters of Earth's distance from the Sun, that alone can't explain the stark difference in environmental conditions.

The Earth's surface temperature averages about 15°C, and ranges from 45°C or so near the equator to -50°C or less at the poles. This moderate range enables water to occur plentifully in all three states – gas, liquid, and solid – and this, in turn, allows a multitude of different habitats, to which life in ever-increasing diversity has adapted. Habitable places on Earth run the gamut from deep ocean hydrothermal vents to rich coastal margins, from savannah to polar deserts.

There are good reasons to suppose that Venus and Earth started out on similar paths. In fact, Venus (like Mars) may have held an advantage in terms of its early biological credentials. Four billion years ago, the Sun was 30% cooler than it is today, making the climate on Venus just about right to harbour oceans and the beginnings of life, while Earth was still a little chilly – especially during the so-called "Snowball Earth" events, when most of the globe was frozen over.

At any rate, both planets most likely had oceans quite early on, and given that life got a foothold so quickly on our own world, there's no reason to suppose it didn't do so at least as quickly on Venus (see figure 16). Indeed, it is reasonable to suggest that life got an earlier start on Venus, given the slightly warmer and more hospitable conditions there compared to those on the young Earth.

**Figure 16** Image of the surface of Venus by the Magellan probe. Notice the segment of a meandering channel, about 200 km long and 2 km wide. These channels are common on the plains of Venus and resemble rivers on Earth in some respects, with meanders, cut-off oxbows, and abandoned channel segments. Most scientists interpret them to have been formed by lava, though some of them could be ancient remnants of riverbeds carved by liquid water at a time when temperatures were moderate and water was still plentiful on the surface of Venus

Photosynthetic microbes likely bathed in the sunlight in warm early oceans on Venus, while other microorganisms tapped the abundant chemical food sources on offer near oceanic hot vents, perhaps even mingling in hot springs on the continents. Given the larger amounts of available energy, higher temperatures, and variable planetary conditions, evolution on Venus might have progressed at a faster pace, and an abundant and diverse biosphere of microscopic organisms could have been present a few hundred million years after the planet's formation. Life probably colonised the subsurface and may have even expanded on to the continental margins in ever-increasing complexity. But then, perhaps half a billion years after Venus formed, the evolution of the two planets began to diverge.

On Earth, the element most crucial to life (as we know it) – carbon – and the all-important biological solvent – water – are continuously recycled. At the heart of this recycling is the slow, steady movement of great pieces of semi-rigid crust over the more plastic layer of the upper mantle, a process known as plate tectonics. Water evaporates from the surface (becoming water vapour), cools, condenses, forms clouds at heights of no more than a few kilometres, and then falls as rain. While in the air, water droplets mix with carbon dioxide to make a weak acid, called carbonic acid.

Back on the ground, this rainwater erodes rocks rich in calcium-silicate minerals. At the same time, the carbonic acid chemically attacks the rocks, releasing calcium and bicarbonate ions which are carried by rivers into the ocean. Plankton and other marine creatures use these ions to build chalky shells of calcium carbonate; then, when they die, they settle on the seabed and their shells form carbonate sediments. As time goes by, the seafloor glides along like a giant conveyor belt, transporting the carbonate sediments to the margins of the continents. Where plates meet, the oceanic plates dive under the continental plates, often with spectacular results. The pushing down of the Pacific plate underneath western North

America, for example, has spawned the volcanically active Cascade Mountains and triggered the 1980 eruption of Mount St. Helens. In other regions, such as that of the San Andreas Fault, the parallel motion of plates can lead to devastating earthquakes.

For those living where great crustal plates slip, slide, and collide it may seem as if the restless motions of Earth's crust are nothing but destructive. Yet, on a global scale, plate tectonics is crucial to life, much as an organ system is to the body.

As oceanic plates, rich in carbonate sediments, descend under the continental plates, their rocks become heated and pressurised. This causes the calcium carbonate in them to react with quartz (a form of silica), re-forming silicate minerals and releasing carbon dioxide gas. The carbon dioxide then re-enters the atmosphere through fissures in the crust – at the mid-ocean ridges or, more violently, during volcanic eruptions.

Plate tectonics operates in conjunction with life to effectively regulate the amount of carbon dioxide in the atmosphere. If carbon dioxide levels rise, photosynthetic organisms thrive and convert more carbon dioxide to oxygen. Also, there's a tendency, over long periods, to sequester more carbon marine organisms into rocks, so that the atmospheric carbon dioxide is kept in check.

But not so on Venus. We don't know for sure whether Venus ever had plate tectonics or whether this crucial recycling mechanism ground to a halt at some early point in the planet's career. What's certain is that today the Venusian crust is unremittingly thick and immobile. Heat and gassy compounds gradually build up underneath it and very occasionally – every 700 million years or so – a cataclysmic event takes place in which all the pent-up energy is suddenly released and most of the planet is paved with fresh lava.

Venus is devastatingly hot because it has a massive carbon dioxide atmosphere that traps the Sun's heat. And it has this stifling atmospheric blanket because it has no way to harmlessly store away much of the carbon that exists in its carbon dioxide greenhouse. The root of the problem seems to be that Venus is just a bit too close to the Sun. The extra warmth associated with this proximity encouraged a high rate of evaporation from the ancient surface waters of the second planet. More and more water vapour was pumped into the atmosphere, and the only place for it to go was upward. Instead of condensing as rain at low altitudes, as happens on Earth, the water vapour on Venus rose to heights of 100 kilometres or more. And that was disastrous, because at those heights the water molecules were smashed apart by solar ultraviolet rays into oxygen and hydrogen atoms. The oxygen atoms combined with sulphur in the atmosphere to form sulphuric acid and sulphur dioxide, while the lighter, faster-moving hydrogen atoms simply escaped into space. As the hydrogen leaked away, so, effectively, did Venus' water. With an end to rainfall, there was no way to wash carbon dioxide out of the atmosphere. The entire carbon cycle shut down, including the horizontal and vertical movement of plates (plate tectonics) that helped to store these minerals within the crust. The greenhouse thermostat was turned up to full and left there.

Disturbingly, our own planet will eventually share Venus' fate. As the Sun ages and grows steadily hotter, the Earth too will warm up and, about one billion years from now, begin to lose its oceans. Its atmosphere will become choked with carbon dioxide and surface temperatures will soar to the point at which most life becomes impossible. Human-induced changes to the environment threaten to bring about Greenhouse Earth on a dramatically shorter timescale.

## Living hell?

Common sense says nothing can live on a planet whose surface is hot enough to melt lead and even zinc. Common sense might also suggest that no life could endure the scalding, superheated water pouring out of deep-sea vents, or an extended plunge in liquid nitrogen at -196°C, or very high levels of acidity, alkalinity, or radiation. But such organisms exist, we know, right here on Earth, and new, more outrageously extremophilic microbes are being discovered every year. Perhaps, given our widening horizons of what's biologically feasible, it's time to look again at the potential habitability of Planet Hell.

A reasonable starting point is to ask whether, in theory, there might still be some liquid water in the rocks underground on Venus that could provide an oasis for life. It's possible to estimate how much heat conducts down through these rocks from the surface, but the results aren't encouraging. They offer no hope of a region remaining in the subsurface where liquid water could be stable. If any underground water persists on Venus it would have to be in a so-called "supercritical" form. A supercritical substance that would exist as a gas under Earth's surface conditions could be kept liquid by the enormous atmospheric pressures on Venus. The trouble is, supercritical water bears little resemblance to ordinary water and its properties suggest it wouldn't interact well with any kind of organism with which we're familiar.

Venus' pressure-cooker environment does favour the idea that another substance – carbon dioxide – might be liquid below the surface, and liquid carbon dioxide is a compound with some interesting properties. Although most terrestrial carbon dioxide takes the form of gas, there is liquid carbon dioxide under Earth's oceans due to the higher pressure there. Organisms on Earth can tolerate it, but these places don't seem particularly friendly to life. For any creature to be able to make use of liquid carbon dioxide, its biochemistry would have to be radically different from the one we know, making further theorising difficult.

If Venus underground seems biologically unpromising, then on the surface it's downright depressing. Where even zinc would collect in molten pools, there's no chance of liquid water surviving. In his 1996 book *Venus Revealed*, David Grinspoon (now curator of astrobiology at the Denver Museum of Nature & Science) suggested that the peaks of some Venusian volcanoes, soaring five kilometres high, might be biologically interesting because of their lower temperatures. These summits, he suggested, might be the place to look for some kind of non-carbon-based lichen that harvests ultraviolet instead of visible light. But, as in the case of hypothetical denizens of liquid carbon dioxide, it's hard to carry such speculation further if the biology is so alien.

With the surface and subsurface ruled out for life as we know it, the only place left to search on Venus is the lower atmosphere. What if microbes had evolved in the ancient oceans of the second planet, billions of years ago, and then, as the surface waters boiled away, adapted to exist in a purely airborne state?

The most interesting region in this respect is Venus' so-called lower cloud deck, at an altitude of about 50 kilometres (30 miles). Some water vapour still lingers here, space probes have revealed, and there are clear signs of chemical disequilibrium, a condition in which different substances are encouraged to react with each other. There are potential nutrients here, too, the temperature (30° to 70°C) is toasty but biologically manageable, and the pressure (1 bar) is like that at Earth's surface. The acidity is high, with a pH level of 0.[56] But that isn't a show-stopper either because there are places on Earth with strikingly similar conditions – place, that are swarming with bacteria.

## Thinking the unthinkable

The Valles Caldera region of New Mexico, northwest of Santa Fe, is the wreckage of an ancient supervolcano. It last exploded over a million years ago but is still the site of much geothermal activity. In the Sulfur Hot Springs area, hydrogen sulphide and other gases bubble out of ponded water that sprawls across a basin bigger than a football field. Basalt cliffs provide a scenic backdrop and pine trees grow up all around. The springs themselves are bare of vegetation but not, it turns out, of other kinds of life.

In the odorous, acidic brew of the Sulfur Hot Springs dwells a thriving community of remarkable organisms. These thermophilic (heat-loving) microbes get along just fine in waters that reach 70° or 80°C, with a pH of 1 or 2. (A pH of 7 is neutral; lower numbers indicate more acidic conditions. Lemon juice, for example, has a pH of about 2, battery acid of about 1.) Surprisingly, the most acidic waters in the Sulfur Hot Springs locale also host the highest concentrations of thermophilic bacteria. This is because although the strong acid doesn't harm the bacteria, it does deter any predators from invading their habitat, leaving them free to multiply unchecked.

To power their cellular machinery, the bugs of the hot springs use hydrogen sulphide, a gas well known for its nauseating smell of rotten eggs. As the hydrogen sulphide wells up from the subsurface, it's oxidised by the bacteria as part of their metabolism, forming sulphuric acid in the process. Venus and New Mexico may literally be worlds apart, but conditions in the Venusian lower cloud deck and those in Sulfur Hot Springs of the Valles Caldera show some striking parallels. The prevailing temperatures are about the same and both environments are rich in hydrogen sulphide. Although liquid water isn't plentiful in the lower atmosphere of Venus, measurements by space probes suggest a reasonable concentration of several hundred parts per million (close to 0.1%). The pH value of 0 in the Venusian clouds is even more extreme than that enjoyed by the hot springs bacteria, but there are known Earth organisms that can handle such super acidity. Some of these were discovered by Christa

Schleper of the Max Planck Institute for Biochemistry in Germany.[57]

The Venusian atmosphere is interesting in another way: it contains, on the one hand, large amounts of hydrogen and hydrogen sulphide, which are reducing substances, and, on the other, plenty of sulphur dioxide and oxygen, which are oxidising. Under normal circumstances these reducing and oxidising gases would quickly eliminate each other, but they are still there, coexisting, which means that some process must be continually producing them.

Earth's atmosphere shows a similar kind of disequilibrium. It contains both oxygen (a product of photosynthesis) and methane (a product of, among other things, flatulent cows, rice paddies, and microbes with an ancient ancestry called the *archaea*). Methane and oxygen react with each other to form carbon dioxide, yet the supply of them in Earth's atmosphere never runs out. To an alien race observing our planet from around some other star, this juxtaposition of methane and oxygen would be a strong indicator of life. While the disequilibrium in Venus' atmosphere isn't as pronounced, it certainly demands an explanation, and one possibility is that biology is at work.

## Cytherean specks

Another clue as to what may be going on came from the Russian Venera probes 9 to 14, which carried out studies of the second planet between 1975 and 1982. Sensors on these probes detected many particles in the lower Venusian cloud deck that are about one micron (one millionth of a metre) across, and elongated on one axis. Tantalisingly, many terrestrial microbes are comparable in size and shape. One of the goals of the Pioneer Venus multiprobe mission in 1978 was to investigate the composition of these particles, which based on data from a Soviet spacecraft might be different on the inside than on the outside. But unfortunately the instrument intended to do this, a high-temperature oven designed to break down substances for analysis, failed to send back any results due to a technical glitch.

Venus is remarkable in many ways, not least because of its rotation about its axis, which is very slow (with a period a little longer than the 225 Earth days it takes Venus to orbit the Sun) and apparently in the opposite sense to every other planet in the Solar System, because it is upside down. Winds on Venus, too, are very odd. At, or close, to ground level they're virtually non-existent because of the planet's even surface temperature (itself a result of the immense greenhouse blanket). But at greater altitudes the winds pick up dramatically. In fact, the bulk of the atmosphere super-rotates, meaning that it circulates much faster than the planet itself spins. Close to the surface the atmospheric rotation period is about 117 Earth days; higher up it drops to as little as 4 to 6 days. As a result, the Venusian atmosphere can exert tremendous shear forces, especially on anything that is rising or falling. Interestingly, the super-rotation on Venus could be a boon for airborne life, because it would boost the potential for photosynthesis, by reducing the continuous time spent in darkness.[57]

The clouds of Venus are much larger, more continuous, and more stable than clouds on Earth, and would make a correspondingly more suitable habitat. Even in Earth's atmosphere, recent research has shown, cloud droplets at high altitude can serve as microenvironments in which bacteria actively grow and reproduce.[58] Birgit Sattler of the University of Innsbruck, Austria, concluded that a key limitation for life in cloud droplets is the residence time in the atmosphere. On this score, the Venusian atmosphere has a huge advantage, with a residence time measured in months (well within typical bacterial reproduction times). Residence times in

Earth's atmosphere are, by contrast, only a few days because particles inside cloud droplets are quickly rained out.

## In Sagan's footsteps

Carl Sagan, astronomer and science communicator extraordinaire, was among the first scientists of the modern generation to revisit the possibility of Venusian biology. So convinced was he that life exists on Venus that, before his untimely death, he planned to found a society on the subject. Unfortunately, his enthusiasm wasn't yet widely shared and no one showed up at the inaugural meeting.

In recent times, scientific consensus has shifted to view life on Venus as a more realistic possibility. Recent academic papers have argued for the existence of life in the Venusian atmosphere and have fleshed out the details of a limited but sustainable microbial ecosystem.[59]

One major obstacle facing airborne life anywhere is ultraviolet and other forms of penetrating radiation, levels of which are much higher than on the ground. Organisms on Earth cope with this problem by using layers of water, dead cells, or specialised pigments as radiation shields. Later in Earth's history, ozone in the upper atmosphere formed an effective ultraviolet barrier behind which more complex organisms could evolve. In the Venusian atmosphere, the scarcity of water and the lack of protective ozone put much greater constraints on life.

There is, however, a way around the radiation problem. It involves elemental sulphur, specifically in the form of a ring-like structure of eight sulphur atoms, known as S8 or cycloocta sulphur. This substance (familiar as the powdery kind of sulphur sold in chemists' shops) is fluorescent: it absorbs ultraviolet and then reradiates the energy at visible wavelengths. A coating of S8, therefore, would not only protect any microbe inside from harmful radiation, it would also give it access to visible light for use in photosynthesis.[60]

What's more, S8 is the most stable sulphur compound known – so stable that even sulphuric acid can't attack it – and, computer modelling suggests, it is present in enriched amounts in the lower cloud deck of Venus. Particles covered by it are the likely cause of the black streaks shown in the reflected ultraviolet light of figure 17.

**Figure 17** Image of the clouds of Venus as seen by the Pioneer Venus Orbiter in the ultraviolet light from 5 February 1979. The dark streaks are produced by absorption of solar ultraviolet radiation and could conceivably be caused by microorganisms that use elemental sulphur as a "sunscreen"

The big issue is, what is the nature of these particles? Are they mere specks of volcanic dust or living cells?

S8 is used by many terrestrial microbes, mostly of very old ancestry. Green sulphur bacteria, and some purple sulphur bacteria and cyanobacteria (formerly known as blue-green algae) are known to deposit sulphur outside their cells. Such primitive creatures are common in Yellowstone National Park and similar locations. They use an ancient photosynthetic pathway, called Photosystem I. Photosystem II is the familiar light-harvesting, oxygen-releasing pathway found in green plants; Photosystem I exploits a different set of compounds. Carbon dioxide and hydrogen sulphide are tapped, using energy from the sun, to produce organic carbon, water, and elemental sulphur. This type of photosynthesis would fit right into the environment of the Venusian lower cloud deck, replete as it is with the necessary ingredients. The reaction would even churn out the elemental sulphur that could form a radiation shield.

An alternative metabolic pathway for Venusian life might be based on chemical energy instead of light. Chemical energy could be harvested with carbon monoxide and sulphur dioxide to form hydrogen sulphide or carbonyl sulphide (consisting of one carbon, one oxygen, and one sulphur atom). Carbon monoxide and sulphur dioxide are plentiful in the Venusian atmosphere and the resulting sulphides could immediately feed back into the photosynthetic pathway. This would close the feedback loop and allow the recycling of chemical nutrients, light being the only outside energy source required. Interestingly, on Earth carbonyl sulphide occurs exclusively in nature as a result of biological activity. On Venus it is the second most abundant sulphur compound in the atmosphere.

Life may have originated in the long-lost oceans of Venus or been seeded there by microbes carried on meteorites from Earth. Green and purple sulphur bacteria would have found a home-away-from-home in the steamy waters of the second planet. As in the case of Mars, the chances of early habitation on Venus, achieved by one means or another, are surprisingly high when all the factors are weighed. When, later on, conditions deteriorated as the runaway greenhouse effect took hold, the residents of the surface ocean would have had only two options – escape to the lower atmosphere or become extinct. Again, given the resilience and adaptability of microbial life on Earth, it's hard to believe that some organisms didn't permanently take to the air.

## From Venus with life

These ideas about possible life in the Venusian atmosphere stirred up plenty of excitement in the scientific community, and may even have helped save the European Venus Express mission from cancellation. Unfortunately, Venus Express wasn't built to go chasing after biology in the lower cloud deck. Only a special-purpose sample-return mission could do that: a spacecraft designed to trap some of the mysterious microbe-sized particles and bring them back to Earth for a close look.

One such possible mission would involve launching a mothership on a minimum-energy (Hohmann) trajectory to our inner neighbour, where it would be captured by aerobraking (using atmospheric drag to slow the spacecraft) and eventually achieve a circular orbit with an altitude of 150 kilometres.[61] A probe, equipped with a collector, would be released, and would descend by parachute to within 50 kilometres of the surface. At this height, a balloon would inflate, allowing the collector to drift leisurely through the atmosphere, gradually gathering its Venusian take-away. Job done, a rocket would fire on the probe, lifting it back up to the mothership for the trip back to Earth. Upon its return, the spacecraft would rendezvous with the International Space Station to await the next shuttle home.

Such a mission could be done quite cheaply, as Venus is closer to us than Mars and no actual landing is called for. Still, given NASA's current focus on a crewed return to the Moon, and beyond that a manned Mars excursion, the chances for a daring new US probe to the second planet seem slim. Help, however, may come from an unexpected direction; the Swedish Space Agency is considering, as its first interplanetary sojourn, a mission to Venus.

# 7

# Hidden Depths

Beyond the orbit of Mars, beyond the main asteroid belt, once feared as a hazard to future space-farers, lies the realm of the gas giants, Jupiter, Saturn, Uranus, and Neptune. Here the Sun's rays are weaker, the intensity of solar radiation falling away quickly as we move away from the Sun. Jupiter, at a little more than five times Earth's distance from the Sun, receives less than one twenty-fifth the amount of solar light and heat on each unit area. At Saturn's distance, the heating and lighting effect is 90 times less than what Earth enjoys.

Not surprisingly, the temperature at the cloud tops of the gas giants makes a cold Arctic night seem sultry: −150°C in Jupiter's case, −180°C in Saturn's. The moons of the planetary behemoths have surfaces that are even cooler, their bright icy surfaces reflecting back much of the Sun's feeble radiation. Jupiter's great moon, Ganymede, largest in the Solar System, shivers at an average temperature of −156°C; Neptune's big companion, Triton, barely reaches 40 degrees above absolute zero (−233°C).

## Far-out speculation

These are not places that look, at first glance, even remotely habitable. Yet that hasn't stopped scientists, over the years, from making a number of ingenious suggestions as to how life might be possible in the seemingly frozen depths of the Solar System. In 1954, the geneticist J. B. S. Haldane, speaking at a symposium on the origin of life, mused about an alternative biochemistry in which water was replaced as a solvent by liquid ammonia. His idea was picked up by the astronomer V. Axel Firsoff, who pointed out that the atmospheres of the gas giants, being plentiful in ammonia, might be natural places to look for this kind of alien biota. In 1976, Carl Sagan and astrophysicist Edwin Salpeter went further, publishing a paper called "Particles, Environments, and Possible Ecologies in the Jovian Atmosphere" in which they weighed the possibility of a vibrant ecosystem in Jupiter's upper cloud deck populated by "sinkers," "floaters," and "hunters," taking the form of giant gasbags analogous to oceanic animal life on Earth. There's certainly organic material in Jupiter's clouds and the planet has its own vast internal heat supply due to slow, steady gravitational contraction. As a result, the temperature of the planet quickly climbs with increasing depth below the chilly topmost layer of the atmosphere.

No one, until quite recently, gave much heed to the biological or biochemical credentials of the satellites of the gas giants, with one exception – Titan, the biggest moon of Saturn (which we'll look at more closely in the next chapter). The reason for this lack of interest was twofold: very low temperatures and, as in the case of our own Moon, an almost complete absence of atmosphere. Titan alone looked interesting because, uniquely among satellites in the Solar System, it is surrounded by a dense blanket of gases. As for the other deep space moons, surely no object, so remote and exposed to the vacuum and radiation of space, could be of interest in the search for life?

## A turn of the tide

If science in general, and astrobiology in particular, has taught us anything over the past forty years or so, it's to keep an open mind. The universe is a continually surprising place, full of unusual events, structures, substances, and, it turns out, potential habitats.

Life as we know it, or can seriously speculate about, has three basic requirements: a source of energy, the building blocks for organic substances, and water, or, in a broader sense, some suitable biological solvent. The moons of the gas giants seem, on the face of it, to be deficient in all three. Their supply of solar energy is feeble, they have no obvious source of organics aside from the occasional meteorite or comet that may collide with them, and any water on the surface is evidently deeply frozen as ice.

Larger, planetary objects, Earth-sized or bigger, have interiors that give off plenty of heat, even now, more than four billion years after they coalesced from the solar nebula, because of the steady decay of radioactive elements. Gas giants, such as Jupiter and Saturn, have an additional source of internal heat in the form of gravitational contraction – the release of energy as the loosely-packed, outer layers of which they're composed slowly pull themselves more tightly together over aeons.

With a few exceptions, a mere satellite, due to its smaller size, has neither of these ways to warm itself from within. Yet it may have another kind of heater if it happens to move in the right kind of orbit around its parent planet. The first person to entertain this idea appears to have been the American astronomer John Lewis who, in a 1971 paper, wrote: "... the Galilean satellites of Jupiter, and the large satellites of Saturn, Uranus, and Neptune, very likely have extensive melted interiors."[62] Lewis realised that a biggish moon in an elliptical orbit would experience a continual stretching and stressing of its interior because of tidal forces. These forces stem from the gravitational interaction between a planet and a moon; on Earth, tides are caused by the influence of both the Moon and the Sun and are most noticeable in the rising and falling of seawater levels on coasts. But tides can also affect solid or semi-solid rock, and this tidal flexing is most pronounced in the case of a big moon moving not too far away from a giant planet in an orbit that is not exactly circular. The changing distance between planet and moon leads to differences in the extent and position of tidal "bulges" raised on the moon, which in turn cause friction and heating in the moon's interior.

Less than a decade after Lewis wrote his paper, the twin Voyager probes, swooping through the Jupiter system *en route* to Saturn and beyond, vindicated his theory in glorious technicolour. One of the Jovian moons, Io, was so wracked by tidal heating that its entire surface was covered in orange, yellow, and red sulphur, and sulphur compounds spewed out of volcanoes. Scientists were astonished to see a number of volcanic eruptions in progress and colossal eruption plumes rising high above the moon's limb.

Io is roughly the size of our own Moon and its distance from Jupiter is only slightly more than that of the Moon from Earth, so the tidal forces acting on it are extreme. At not quite double the distance from Jupiter is another of the four big Galilean moons (so named because Galileo Galilei was the first to see them), Europa. And its story, though very different than Io's, is no less intriguing.

## "Attempt no landings there"

Europa takes centre stage in *2010*, the sequel to Arthur C. Clarke's famous novel (and Kubrick's brilliant film), *2001: A Space Odyssey*. Extrapolating from the theoretical work on tidal heating started by Lewis, Clarke saw Europa as the likeliest abode for extraterrestrial life in the outer Solar System and made it the focal point for a new genesis triggered by the mysterious black monoliths. In the closing scenes of *2010*, as Jupiter blazes forth as a new star, the monolith aliens send humanity an enigmatic message: "All these worlds are yours except Europa. Attempt no landings there." In fact, landing a probe on Europa is exactly what scientists would love to do, because this strange world – a little smaller than the Moon, at a diameter of 3,160 kilometres (1,950 miles) – has captured the interest and imagination of every astrobiologist.

Seen close up, first by the passing Voyager probes and then in immensely greater detail by the Galileo spacecraft during its circumnavigations of Jupiter between 1995 and 2003, there is no mistaking Europa's oddness. In general, its surface resembles a cracked and stained eggshell. It is the smoothest large rocky body in the Solar System with no feature rising more than a few hundred metres above its surroundings (see figure 18). Everywhere its icy surface is crisscrossed by great fissures and bucklings of ice – the visible signs of the tremendous tidal forces at work on this world. Although these forces are not as severe as they are on Io, still they are strong enough to cause very significant, continuous tectonic activity.

After Io, Europa is the most geologically dynamic moon in the Sun's domain. Yet we see only modest signs of that activity at the surface. The cracked shell of water ice hides much of what is happening internally. But through a combination of theory and observation, researchers have gradually peeled away the mystery of what may be happening in the unseen depths of this moon.

Years before the Galileo spacecraft arrived in orbit around Jupiter, scientists had surmised that Europa might have an ocean – not a terrestrial-type body of water in plain view, but a great underground ocean entirely sealed off by the icy crust. The real action on Europa, researchers guessed, was happening not at surface level, but on the stygian bed of this global sea. The big news in astrobiology

**Figure 18** Europa, a moon of Jupiter, appearing as a thick crescent. Image taken by the Galileo spacecraft

was that Europa might have underwater hydrothermal vents like those on Earth, newly-found to be havens of weird, heat-loving life (see figure 19).

Observations by the Galileo probe indicated that Europa's sub-surface has very similar properties to salt water, which lent support to the notion that Europa has a liquid interior beneath its icy shell. Seen at close range, the surface cracks on the moon looked very much like those associated with mobile icebergs driven by subsurface water on Earth. A picture of Europa's interior was built up in which a dense metal core and rocky mantle is surrounded by a low-density ice crust. The zone of liquid water beneath the crust,

**Figure 19** Tube worms, limpets and protists (single-cell organisms) at the South-East Caldera of the Axial Volcano on the Juan de Fuca Ridge (NE Pacific). The photo of this hydrothermal vent was taken from a submarine on 29 August 2008

measurements suggested, might be an astounding 80 kilometres deep. If correct, this estimate would give Europa three to four times more water than all of Earth's oceans put together.

A major unknown is the thickness of the ice crust overlying Europa's ocean; estimates range from several hundred metres to tens of kilometres, with the truth probably lying somewhere in between. In areas that have been free of volcanic activity for a long time the thickness of the ice crust must be at least a few kilometres; in other places that have seen more recent activity the ice is likely to be much thinner (see figure 20).

In any event, the existence of this great body of water so far from the Sun is an unexpected bonus for astrobiologists. What life, if any, might lurk in its depths?

## The submariners of Europa

Most oceanic life on Earth, like most land-dwelling fauna, depends ultimately on sunlight. Photosynthetic organisms, many of them very small, floating or swimming near the surface, form the basis of the food chain that supports almost every other creature living in the sea, from shrimps to sharks. But if there is life in Europa's ocean it must be underpinned primarily by a different source of energy because virtually no light can filter down through the icy crust to the waters below.

Chemical energy is an obvious alternative to light, given the success of hydrothermal vent communities living in almost complete darkness on our world. Another possibility is that life-forms on Europa tap thermal energy – the heat generated inside the moon – a source that (as far as we know) no terrestrial organism exploits. Thermal energy may have been important to early terrestrial life before photosynthesis took over, as suggested by Anthony Muller,

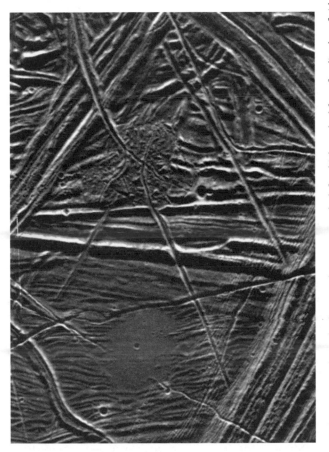

**Figure 20** Close-up of Europa's surface. The flat smooth area seen on the left, which resulted from flooding by a fluid, is about 3.2 km (2 miles) across and may suggest closer access to liquid water with a possible biosphere below. The smooth area contrasts with a distinctly rugged patch of terrain farther east, to the right of the prominent ridge system running down the middle of the picture.

formerly at Washington State University.[63] Muller proposed the term "thermosynthesis" to describe this hypothetical mode of powering an organism's metabolism. If his idea is correct, it may be that life on Europa, deprived of a light-harvesting option, has remained stuck at the thermosynthetic stage.

Photosynthesis clearly has great advantages: sunlight is predictable and comes in a highly ordered state. Thermal energy, even if potentially rich in quantity, is more random and subject to fluctuations. The reason for this goes back to a basic tenet of physics, called the second law of thermodynamics, which insists that any closed system will become more disorderly over time. Plants obey this law by taking up light energy in a highly ordered state (light waves with a specific frequency) and radiating heat across a broad spectrum that is more chaotic. The same is true of every life-form. The problem for an organism is that if it were to feed on relatively disordered thermal energy, it would have its work cut out to extract enough for its metabolic needs before having to excrete waste energy that was even less ordered.

Convection offers one possible solution for thermosynthetic organisms. This familiar process is seen in action when water boils in a pan: the heated water rises and allows colder water to be drawn toward the heat source. The resulting circulation of liquid carries kinetic energy, or energy of motion. In some natural circumstances, this energy could be harvested directly, for example by an organism attached to a rock with an elongated appendage exposed to the moving water. The transfer from kinetic energy to stored biochemical energy would work essentially like charging a battery; all that would be needed is a slight chemical gradient, induced by the moving water, and time to charge the system enough to build energy- storing chemical compounds, such as ATP (adenosine triphosphate) or its equivalent.[64]

Life on Europa might also be able to make use of chemical gradients directly. Saltwater is denser than freshwater and there are many natural environments on Earth where a layer of saltwater underlies fresh. The transition between the two over a short distance sets up a chemical gradient, with many salt molecules on one side of the transition zone and few on the other. If an organism could move between these layers it could harvest energy available in the gradient. Calculations show that by expelling three water molecules in the high salt layer, an organism could drive the assembly of one ATP molecule from its precursor, ADP (adenosine diphosphate). Simply by rising back up to the freshwater layer, the creature could then restock its supply of water before beginning the cycle again. Many species on Earth, from cyanobacteria to complex animals such as sharks, can withstand strong chemical gradients, so this adaptation is by no means far-fetched.[64]

The possibility that something larger than microbes might inhabit Europa's ocean fires the imagination. But it hinges critically on how much food is on offer in those alien depths. A 2003 paper took an ecosystem approach to possible life on the enigmatic moon.[65] The authors estimated how much biomass could be supported by the nutrients assumed to be present in Europa's subsurface ocean. They also assumed the presence of different types of organism that would feed on the original producers (comparable to plants on Earth). Their conclusion: Europa's ocean should have enough nutrients to support an ecosystem with something like shrimp-sized animals as the top predators. Future observations will tell if the estimates they made were anywhere near the mark, but this was the first serious attempt to evaluate the prospects for multicellular life on Europa. Unfortunately, it led to wild speculation in some quarters that the Jovian moon might be home to all manner of species up to and including great whale-like leviathans.

Little is known about the chemistry and environmental conditions of the Europan ocean. Surface colourations of the ice and results from modelling the history of the moon suggest that it could be extremely salty. Some researchers have suggested that the ocean may be very acidic due to a high concentration of sulphuric acid – sulphur compounds having been detected on the surface (though these may derive from volcanic eruptions on neighbouring Io). Others believe the underground waters may be very alkaline because of a high ammonia content. Only close-up studies by future spacecraft will shed further light on these matters.

## Children of the icy giants

Europa was the first moon to be suspected of having a great body of water beneath its icy crust. But astrobiologists quickly began to cast their net wider. What of the other big moons of the gas giants? And what of such moons – presumably not uncommon – that might accompany large planets around other stars? Suddenly, unexpectedly, the scope for extraterrestrial life seemed to be greatly extended both within the Sun's realm and beyond.

The other two Galilean moons, Ganymede and Callisto, move around Jupiter in larger orbits than those of Io or Europa, so the amount of tidal heating they experience is quite a bit less. This is reflected in the older, less dynamically-active appearance of their surfaces. Callisto is one of the most heavily cratered objects in the Solar System, a sure sign that its surface has been little altered over the past four billion years. Ganymede, however, is a more interesting place, with a history of some geological and tectonic change and the real possibility of a subsurface ocean.

Ganymede narrowly beats out Titan as the biggest moon in the Solar System, with a diameter of 5,268 kilometres; it's also the only satellite to have its own magnetic field, providing some protection from Jupiter's radiation blitz. Its size is important, as it implies a large core, with sufficient remaining stock of radioactive elements to provide some internal heat in addition to that which is tidally generated. The moon's surface is mainly a mixture of rock and water ice, with a dash of frozen carbon dioxide ("dry ice"), and ranges in temperature from −130°C to −170°C. Two distinct landscapes are apparent: a dark-cratered region with numerous in-filled craters, and a more youthful, light-coloured terrain, cracked and grooved. The light-coloured terrain is interpreted to be the result of resurfacing events when liquid water from further down re-paved the surface (see figure 21). Most scientists implicate a deep subsurface ocean as the source of the resurfacing event, but pockets of liquid water could also exist much closer to the surface.

Based on observations from the Galileo probe and the morphology of the ice, the ice crust is thought to be between 80 and 150 kilometres thick. In general, whatever can be said about the prospects for life on Europa applies also to Ganymede, though gaining access to the larger moon's subsurface ocean to prove any assertions about it is a daunting task. The best that can be hoped for in the foreseeable future is that spacecraft may detect traces of biology or organic material that have worked their way up from the ocean to the top metre or so of the ice crust.

Another moon of great astrobiological interest is Enceladus, which orbits within the tenuous, outermost ring of Saturn. Although barely 500 kilometres in diameter, it has been the focus of much attention ever since the Cassini spacecraft, in its travels around Saturn, spotted a plume of water vapour venting from the moon's south polar region.

Enceladus sports a variety of smooth, youthful-looking terrains, pointing clearly to the fact that its surface is changing even today (see figure 22). This discovery, together with that of the vented material, came as a major surprise, given the moon's diminutive size. Evidently, Enceladus, like Europa and probably Ganymede, has liquid water beneath its crust, and with that the potential for life.

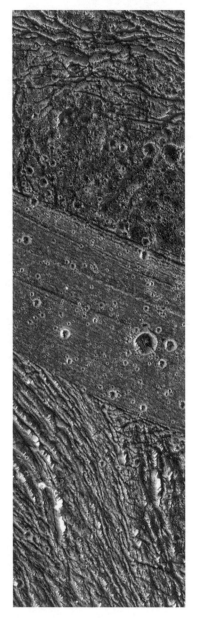

**Figure 21** Close-up image of the surface of Ganymede. Notice the bright terrain (called Arbela Sulcus) slicing north–south across the older grooved terrain. The bright terrain likely formed when liquid water from below repaved part of the surface.

**Figure 22** Enceladus, as imaged by the Cassini Orbiter on 14 July 2005. Even though Enceladus is only 490 km in diameter, it exhibits a bizarre mixture of softened craters and fractured terrain, indicating a molten interior

Analysis of the vented material, made possible when Cassini swooped through the ejecta, suggests that it comes from a body of subsurface water and contains carbon dioxide, methane, possibly nitrogen, and trace quantities of organic compounds. This treasure-trove of sampled compounds, possibly deriving from the deep and heated interior of the moon, catapulted Enceladus from a no-hoper to one of the front-runners in the search for extraterrestrial life. The heat needed to keep the water liquid is probably supplied by tidal interactions between Enceladus, Saturn, Titan, and some of the other moons of the sixth planet.

Titan itself is also likely to have a subterranean ocean. According to some suggestions this ocean might be rich in ammonia, which would act as an antifreeze and allow the water to remain liquid at much shallower depths. Given Titan's thick atmosphere and potential for complex organic chemistry, there is even the possibility of Titanian surface life.

Faraway Triton – Neptune's only big satellite – is another candidate for an underground sea. A likely exile of the Kuiper Belt, from beyond the orbit of Pluto, Triton is so dense that it almost certainly has a metallic core and a large rocky mantle harbouring radiogenic elements. Long-lasting geysers have been observed on Triton's surface, which is mostly composed of nitrogen ice. With temperatures diving below −200°C, there is no colder place in the Solar System. Snows of nitrogen, methane, and hydrocarbons form on Triton, and computer modelling suggests that over the moon's lifetime some six metres (about 18 feet) of hydrocarbon sediments have built up on the surface. Yet the cold icy crust insulates the deeper parts of Triton well and a liquid watery ocean, 350 to 400 kilometres deep, is thought to exist 100 kilometres or so under the frozen crust. If ammonia is prevalent in that body, there may even be liquid water a mere 30 kilometres below the surface.

The biggest moons of the biggest planets are the likeliest havens for interior seas. But the case of Enceladus suggests that more of the smaller icy satellites of the Solar System, some of which are well endowed with organic chemicals, also have surprises in store. In fact, it's starting to look as if ice-covered oceans may be much more plentiful in the universe than exposed bodies of water such as those on Earth. And if life can originate and persist in these new-found habitats, we can expect it too to be more common and diverse than previously supposed.

Although many scientists favour an origin of life at hydrothermal vents (such as those anticipated on Europa), it may also be that ice can nurture the beginnings of biology. Ice tends to accumulate organic compounds in liquid brine channels under subzero temperatures, and many life-essential chemicals react quite favourably under cold conditions. Also, microbial activity has been measured down to temperatures of −40°C and some biochemical pathways involving enzyme activity can still operate at temperatures lower than −100°C.[66]

From our experience on Earth we know that life can thrive in association with ice. It constitutes one of the largest habitats for life, supporting a variety of organisms including bacteria, algae, and animals. Life also flourishes under the Antarctic ice shelf and in sub-ice lakes in Antarctica. Even our own planet has endured so-called "Snowball Earth events, when all or most of the surface was frozen over. During these periods, environmental conditions were not so different from those found on icy satellites and yet, as our species can testify, life persevered. Thus, the perception that cold icy conditions are not conducive to life is based on our view-point as warm-blooded mammals. Scientific evidence suggests otherwise.

## Ice drills and budget cuts

There's been no shortage of proposals for missions to orbit, land on, or explore beneath the surface of the astrobiologically-interesting moons of the outer planets. But, so far, none of the space agencies has been willing to follow through on any of these ideas.

A few years ago, NASA began preliminary planning for an exciting project called JIMO (Jupiter Icy Moons Explorer), a nuclear-powered spacecraft designed to orbit three of the Galilean moons – Callisto, Ganymede, and Europa – in turn, exploring each in unprecedented detail over long periods. JIMO would have helped settle important questions, such as whether these bodies do indeed have subsurface oceans. It would also have been able to map the location of organic compounds and other chemicals of biological interest, and determine the thicknesses of ice layers, with emphasis on locating potential future landing sites. A basketball-sized probe would have been released for a soft touchdown on the frozen wastes of Europa.

But the project was scrapped before it ever left the drawing board, just one of the victims of the decision to refocus the American space agency on a manned return to the Moon. Researchers arriving at the 2006 Astrobiology Science Conference in Washington, D.C. were in a sombre mood. JIMO had been ditched and NASA Headquarters had just halved its astrobiology budget, which meant there would be no grants awarded in the field in the coming year. All new proposals and solicitations had been cancelled or deferred, and many scientists who had begun to think of themselves as full-time astrobiologists were left feeling anxious. They were on "soft" money, meaning that their salaries and research expenses were solely grant-supported. With such funds suddenly gone from astrobiology, researchers would have to leave the field and focus their efforts elsewhere. Students who'd been planning their careers around the search for life beyond Earth would be forced out too, and the next generation of astrobiologists would be lost.

After the Galileo and Cassini spacecraft had completed their work around Jupiter and Saturn, respectively, it seemed there would be no new flagship mission to take a closer look at worlds suspected of harbouring great oceans and even life. In particular, a spacecraft able to scrutinise Europa for weeks or months on end would be sorely missed. This moon, and others in the outer Solar System, are so distant that virtually nothing of their internal nature can be learned without a robotic presence nearby. And to have a reasonable chance of making a biological discovery, a probe would need to land on Europa and drill at least a metre below the surface – the minimum depth at which Jupiter's destructive radiation would not tear apart organic or biogenic material that had made its way up from the waters below.

Ideally, scientists would like to send a probe all the way down into the underground ocean on Europa, and some preliminary work has been done on equipment that would have this capability. Most attention has been paid to a mission involving a pencil-shaped "cryobot" that would descend through the icy crust by melting its way along before releasing a "hydrobot", or miniature instrumented sub, for exploring the hidden watery deep.

Such an ambitious robot expedition surely lies decades in the future. But if there is life on Europa, there's just a chance we might detect early signs of it with a more modest mission. Brad Dalton, from NASA Ames, has looked closely at spectroscopic details of light reflected from the surface of the moon. Europa has a lot of reddish-brown patches and he compared the light from these with the spectral properties of known substances on Earth. Some previous observations had suggested that the coloured patches might be organic in nature, possibly consisting of materials that had welled up from the ocean below. What Dalton found was intriguing: some of the spectral features in these areas were similar to the signatures of some known terrestrial bacteria. Of course, such a resemblance is not proof of extraterrestrial life. There are too many uncertainties involved: possible interferences, minerals, ice, unknown components that might camouflage or exhibit similar patterns. But if a spacecraft could be set down on those mysterious discolorations on Europa, what would it find?

## A new hope

A major new outer Solar System flagship mission was proposed and given preliminary approval at a joint meeting of NASA and the European Space Agency (ESA) in 2009. Called the Europa Jupiter System Mission (EJSM), it would launch sometime around 2020 and be the first big spacecraft to the outer Solar System following Cassini.

EJSM would consist of two main probes: the NASA-led Jupiter-Europa Orbiter, and the ESA-led Jupiter-Ganymede Orbiter. These probes are designed to carry out a choreographed exploration of the Jovian system before settling into orbit around Europa and Ganymede, respectively. Plans call for each of them to carry at least ten complementary science instruments to monitor fast-changing phenomena, such as Io's volcanoes and Jupiter's atmosphere, map the magnetic field of Jupiter and its interactions with Europa, Ganymede, Io, and Callisto, determine abundances and distributions of surface materials, and characterise water oceans beneath the ice shells of Europa and Ganymede. The overarching scientific goal is to assess more accurately whether the Jupiter system harbours worlds suitable for supporting life.

One of the problems facing any mission to Jupiter is the planet's brutally harsh radiation environment. So extreme is this that if you were to land on Europa, for example, and take off your space helmet, you'd die from radiation exposure before you would from lack of oxygen. Designing and building instruments able to work reliably for long periods under such intense bombardment is a real challenge.

The Europa Jupiter System Mission offers the first chance for an in-depth exploration of Ganymede, the biggest of the icy satellites and one of the more promising candidates for life, after Europa. Although the new mission isn't intended actually to detect life – a follow-up project would have this as its goal – it will be able to assess habitability.

Unlike the cancelled JIMO, the Europa Jupiter System Mission doesn't, at present, include a lander. But a modest lander hasn't been ruled out and hopefully will be incorporated down the road. It would be of key importance in testing surface and subsurface materials on Europa or Ganymede for the presence of biosignatures. If we've learned anything in astrobiology, it's that getting up close and personal is crucial in the hunt for extraterrestrial life.

# 8

# On Titan

In many ways, Titan (see figure 23) is the most unusual and exotic world we know. It is by far the largest moon of Saturn and, for many years, was thought to be the largest moon in the Solar System, though Ganymede of Jupiter has now replaced it at the top of that league. It is also the only moon in the Sun's kingdom to have a dense atmosphere; in fact, the surface pressure on Titan, at 1.5 bars, is 50% greater than on our own world. And Titan's atmosphere is the only one known, other than Earth's, of which the main ingredient is nitrogen (about 95%), the rest being mostly methane with an eclectic sprinkling of organic chemicals. A similar atmospheric composition is believed to have existed on Earth around the time terrestrial life began, and has led to speculation over the years that Titan may be in an early and possibly arrested stage of pre-biological development.

The surface of Saturn's giant moon was a complete mystery until quite recently, obscured as it is by a permanent haze of orange smog. But our knowledge of Titan expanded dramatically with the arrival of the Cassini spacecraft in June 2004, and the extraordinary success of the Huygens probe, which was released by Cassini and parachuted safely down on to Titan's surface. For two hours, Huygens

**Figure 23** Titan, as imaged by the Cassini orbiter. The 1,700 km wide bright region in the centre of the image is called Adiri and lies within the equatorial dune deserts of Titan (greyish areas). The Huygen's probe landing site was east of Adiri

beamed back remarkable images and other scientific data from its landing spot amid a wasteland of ice and frozen hydrocarbons over a billion kilometres away (see figure 24).

## New world

It doesn't rain water on Titan. Images from Chile's Very Large Telescope and Hawaii's Keck Observatory in 2005 and 2006 showed for the first time nearly global cloud cover at high elevations on the moon, from which a persistent morning drizzle fell over the western foothills of Titan's major continent, Xanadu. It was a drizzle of methane.

On our planet we're used to thinking of methane as a gas, but on Titan, because of the low temperatures, this simplest of hydrocarbons is the main *liquid* present. In place of a terrestrial water cycle, Titan has a methane cycle. But the methane that rains down from Titan's skies rarely reaches ground level. Instead, it usually evaporates above the surface, just as rainwater often does over the southwestern desert of the United States. Only very occasionally, it seems, does Titan experience a real downpour. According to Ralph Lorenz of the Applied Physics Laboratory at Johns Hopkins University, these methane monsoons occur maybe once every few centuries and last for months.

Some features on Titan look strikingly Earth-like but are the work of a distinctly non-terrestrial chemistry and weather system. Large parts of the big moon are marked by what appear to be river channels and their tributaries which, though probably dry much of the time, must have been carved by running methane. Certain latitudes on Titan are dominated by dune fields; others, around the poles, are pockmarked by numerous lakes, large and small, filled with liquid hydrocarbons – primarily methane and ethane (the second simplest hydrocarbon) with some nitrogen mixed in. Steep valleys and towering cliffs in many places make the moon's landscape rugged, and slick ice would add to a visitor's perils. By contrast, the Huygens probe landing site was relatively flat and desert-like, its terrain broken only by a plethora of small, ice cobbles (see figure 24).

With temperatures at ground level hovering around −180°C, the prospects for liquid water on or near the surface of Titan might seem non-existent. Yet there is some evidence for running water in the form of bright flows, quite unlike the dark ones of liquid

**Figure 24** The surface of Titan at the landing site of the Huygen's probe on 14 January 2005

hydrocarbons. Low-temperature volcanic activity, or cryovolcanism, powered by tidal forces from Saturn, is the likeliest explanation for once-moving water (see figure 25).

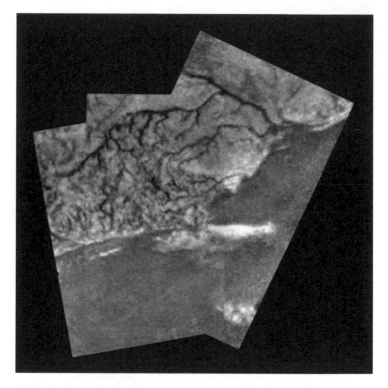

**Figure 25** Drainage pattern of what appears to be liquid hydrocarbons flowing into a lake or ocean on Titan

Activity of a different kind takes place in Titan's dense atmosphere, driven by ultraviolet radiation from the Sun and, to an even greater degree, by radiation accelerated in Saturn's powerful magnetic field. The high-energy bombardment promotes the assembly of organic chemicals by splitting many of the compounds in the atmosphere into radicals – molecular fragments that have a free electron and are very reactive. One of the organic materials made in this way is acetylene, an energy-rich hydrocarbon compound containing two carbon atoms, joined by a triple bond, and two hydrogen atoms. Acetylene forms in Titan's atmosphere as solid particles, which then settle on to the moon's surface. Under normal conditions on Earth, acetylene is explosive and has to be handled with great care. However, in Titan's cold surface environment, in which reactions would otherwise be painfully slow, it's an ideal substance for promoting the build up of complex molecules.

## Titan's garden

Scientists have just begun to take seriously the possibility of life on Titan, either on or below the surface (see figure 26). Speculation about an internal ocean, similar in nature to those suspected on some of Jupiter's moons, has been given a boost by recent data from Cassini. If Titan does have an ocean, its waters are likely to contain a hefty dose of ammonia, thanks to the high abundance of nitrogen on the moon. This would serve as an antifreeze, keeping the ocean liquid at much lower temperatures than if it were pure water – down to nearly −100°C.

In general, it seems that if life does exist on Titan, it must be very different from anything we know on Earth. A combination of extreme cold and strange, hydrocarbon-based chemistry would virtually ensure a unique set of adaptations, from the biochemical level on up. Moreover, a separate origin of life would be almost guaranteed because of the extreme unlikelihood of organisms being transported on meteorites between the inner and outer Solar System. The discovery of "weird life" and a second genesis on Titan would have far-reaching implications for the abundance and diversity of life throughout the universe.

Life on Saturn's big moon would be alien in many respects, yet it might still use some of the same tricks that terrestrial animals and plants turn to in tough times. Take, for example, eastern skunk

**Figure 26** This Cassini radar image shows a big island in the middle of one of the larger lakes on Titan. The island is the size of the Big Island of Hawaii (~ 90 x 150 km). Further down the image, several smaller lakes are seen that seem to be controlled by local topography. The lakes are thought to be filled with liquid methane and ethane plus some nitrogen

cabbage. *Symplocarpus foetidus*, to give it its scientific name, is a strange plant with smelly leaves which belongs to a small group of thermogenic or "heat-generating" species. Remarkably, it produces enough heat to lift its temperature 15°C to 35°C above that of the surrounding air, enabling it to melt its way through frozen ground before flowering. A talent like this, adapted to work at even lower temperatures, would be ideal on Titan, where a creature might use chemistry to make its immediate environment more habitable.

In fact, many types of Earth organisms living in frozen surroundings, even algae, give off some waste heat from their metabolic reactions that can be used to melt their own little watering holes in ice. Such holes are commonly seen on glaciers and in Arctic pack ice and are known as cryoconites. Some of the meltwater underneath the glacial ice sheets even releases methane. Could such a mechanism explain both the apparent smoothness and youthfulness of Titan's surface and the high methane levels in the atmosphere?

Titan's methane has always puzzled scientists. In the moon's lower atmosphere the concentration reaches about 5%, yet there is no obvious source for so much of this gas. The traditional explanation is that a lot of methane became trapped in Titan's icy crust over the aeons and is now being gradually released.[67] But it's hard to understand, in purely chemical terms, how this vast store of methane could have built up, given that the gas is continually broken down in the moon's atmosphere (by a process similar to, but slower than, that occurring on Mars). On Earth, methane is usually an end product of metabolism by microbial organisms, and the same, as we've seen, may well be true on Mars.

One possibility is that organisms in Titan's icy crust free the high-grade energy stored in acetylene by catalysing its reaction with hydrogen, another constituent of the moon's atmosphere, to form methane. Each gram of acetylene exploited in this way could supply about 3.85 kilojoules (nearly 1000 calories) of energy. Given the current level of methane in the atmosphere, and the rate of destruction of the gas, calculations show that about $10^{21}$ kilojoules (one billion trillion kilojoules) would have been released by acetylene-hydrogen reactions over the past ten million years.

One of the few life-forms on Earth for which the amount of energy needed to make new individuals is reasonably well known is the yeast *Saccharomyces cerevisiae*. Making a gram of this yeast "from scratch" takes about 3.2 kilojoules. Assuming that any microbes on Titan use a similar amount of energy to build their bodies, and that all of Titan's methane is produced biologically, it's possible to work out the annual turnover rate and the average concentration of living matter on Saturn's largest moon. The figure that emerges corresponds to just over ten million terrestrial-sized organisms per millilitre – on a par with what would be expected of a slightly nutrient-poor environment on Earth. The density of living material almost certainly wouldn't be uniform over the whole moon, but would vary according to local availability of acetylene and other nutrients.

Based on these calculations, an assessment can be made on whether biological heating – thermogenesis – is a viable option on Titan. The energy-making acetylene reaction mentioned above would yield about 3.85 kilojoules per gram. If organisms on Titan had the same energy needs as *S. cerevisiae*, they'd have about 0.65 kilojoule per gram left over which could be expelled in the form of heat. This, in turn, could melt a substantial amount of ammonia-water ice – close to 100 million metric tons per year.[68]

Of course, all these calculations and estimates have wide margins of error and are, by necessity, based on assumptions about Earth-like microbiology and biochemistry. But, in fact, these assumptions are conservative because anything living on Titan would be far better adapted to local conditions than any terrestrial microbe – for example, it would probably be very efficient in its use of energy.

Whatever the details, a number of key facts come out of this analysis: biothermal heating on Titan is a distinct possibility; the proposed reaction with acetylene is a reasonable metabolic pathway; and the production of methane, if biological, suggests the existence of a large population of organisms on the moon.

## The aliens of Pitch Lake

Astrobiologists are always on the look out for analogue sites. These are places on Earth that have some features in common with extraterrestrial locations. Some parts of the Atacama Desert and Antarctica, for example, as we've seen, are Mars-like. The question is whether any analogue site exists for Titan.

It turns out there are two of them, one in Trinidad, the other in Venezuela. Both are natural lakes of asphalt – a thick liquid mixture of hydrocarbons.

The Trinidadian site, known locally as Pitch Lake, has been used as a source of asphalt worldwide. Large parts of New York were paved with material from the lake. As visitors discover, the asphalt is mostly firm, like congealing lava, and only the active, upwelling areas are truly in a liquid state (see figures 27a and 27b). Occasionally unwary trekkers get stuck in the soft spots and have to be rescued.

Scientists wanting to take samples from the liquid areas of Pitch Lake have to contend with the very gooey nature of the asphalt. It quickly and persistently sticks to skin and anything else with which it comes into contact. The temperature of the liquid patches is about 30°C – typically a few degrees warmer than the surrounding air – demonstrating clearly that there is an underground heat source.

Gas bubbling out of the asphalt contains methane in low but significant concentrations. This and other hydrocarbons have presumably made their way into Pitch Lake from offshore oil reservoirs.

**Figure 27** Pitch Lake in Trinidad & Tobago; (a) overview look, (b) sampling one of the soft spots where liquid asphalt is seeping toward the surface

Could there conceivably be anything alive in such a dark, syrupy brew? Indeed, DNA sequencing and other methods of molecular analysis have confirmed that many microbes make a living in this extraordinary environment, most of them producing methane as a by-product of their metabolism. By an astounding coincidence, the concentration of biological matter turns out to be roughly ten million organisms per milligram of asphalt – in the same ballpark as our early estimate for the density of life on Titan.

Even more remarkable is that some microbes live in Pitch Lake in the virtual absence of water. There's a quantity in science known as water activity, which tells how much "free water" is available for use in chemical reactions or by living things; it excludes any water that's already tied up with other substances or on surfaces. In pure, fresh-water the water activity is 1.0. No organisms have been confirmed to date in an environment where the water activity is less than 0.6. However, when a sample of the contents of Pitch Lake were tested as part of a Discovery Channel documentary, a water activity of 0.49 was found in regions of the lake from which microbial samples for DNA analysis had previously been taken. If this result is upheld by other methods, it would mean that some terrestrial life can metabolise in an almost waterless environment and, just possibly, may not use water as its biological solvent at all, but a mixture of hydrocarbons.

Pitch Lake is certainly not an exact analogue of Titan, most obviously with regard to temperature. However, the Trinidadian asphalt and the organisms found in it show that life can adapt to being permanently bathed in various oils and will hopefully teach us in the long run how their evolution came about. Future research will also show what mechanisms and biochemical changes these organisms employ, and whether perhaps their adaptations would be useful to survival on Saturn's giant moon.

## Cool life

The reaction of acetylene with hydrogen to produce methane isn't the only one that could supply enough energy for a metabolic pathway on Titan. Chris McKay from NASA Ames and Heather Smith from the International Space University in Strasbourg, France, have suggested that heavier hydrocarbons might also serve as a raw fuel for life.[69]

Another possibility involves high-energy radicals. Carbohydrate and nitrogen radicals, for example, could react to form the compounds hydrocyanic acid and cyanamide.[70] On Earth, radical reactions are rarely used in metabolism because they're so difficult to control and can cause internal damage to organisms, ripping cells apart like miniature explosives. But in the super-cold conditions on Titan, these reactions would proceed much more slowly and be biologically manageable. What's more, hydrocyanic acid and cyanamide are thought to have been important ingredients in the origin of life on Earth. Hydrocyanic acid, consisting of one hydrogen, one carbon, and one nitrogen atom, forms amino acids when exposed to ultraviolet radiation. Cyanamide is a compound used in terrestrial biochemistry to help bind amino acids together to form proteins.

Differences in biochemistry suggest differences also in the makeup of cells. On Earth, the membrane of a cell, through which a simple organism interacts with its environment, consists basically of a double layer of lipid (fat-like) molecules. The orientation of these molecules is quite specific, due to the fact that water – the solvent of life on our world – is polar. In other words, a water molecule has a slight positive charge at one end and a slight negative charge at the other. The lipid molecules in terrestrial cells each have a "water-loving" polar head which points outwards, and a "water-hating" non-polar tail which points inwards. Only because of this arrangement can life-as-we-know-it interact with water and extract nutrients from it. If life on Titan uses hydrocarbons, such as methane and ethane, as its solvent of choice, then its cellular architecture must be reversed or non-polar at both ends. Most hydrocarbons are non-polar so that for an organism to interact with them successfully it would need the non-polar ends of its cell membrane molecules pointing to the outside.

## Silicon and life

Cell membranes on Titan might be unusual in another way – by incorporating silicon compounds called silanes in their construction. Silane itself (the silicon analogue of methane) contains one silicon and four hydrogen atoms in a tetrahedral structure, and is a reactive gas under normal Earth conditions. For example, it will immediately combine with water and oxygen to form silicon dioxide, the substance of which sand and quartz are made. However, the environmental circumstances on Titan are vastly different. Oxygen and liquid water are nearly absent, and even carbon dioxide is very rare. Francois Raulin from the University of Paris has suggested that life on Titan might compensate for the lack of available oxygen by using nitrogen analogues in its biochemistry.

Silanes have a number of properties that would make them biologically attractive in the right setting. Silane itself would be a liquid under Titan surface conditions, but polysilanes, consisting of many silane molecules stacked together, would form solids at those temperatures. William Bains from the University of Cambridge Institute of Biotechnology has even come up with a model for photosynthesis based on silicon biochemistry.[71]

Given the abundance of carbon on Titan and carbon's unparalleled ability to form and break bonds in complex substances, it's unlikely that actual silicon organisms exist on the big moon. Silicon-based life is a staple of science fiction. But although Titan probably lacks anything like the rock-boring Horta, featured in one of the original Star Trek episodes, it may well find novel biochemical uses for this element.

## Strangely familiar

The size of living cells on Earth may be a misleading model for life in an environment such as on Titan, which is not dominated by water, a highly polar and relatively high-temperature solvent. As pointed out by Louis Irwin from the University of Texas at El Paso, in an extremely cold, non-polar, liquid hydrocarbon environment, it's possible that life consists of much larger cells, perhaps even visible to the unaided eye. Given the cold surface temperatures, metabolism may proceed very slowly, however. The lifespan of humans and most other organisms on Earth is less than a century, but there is no reason why lifespans on Titan might not be measured in thousands or hundreds of thousands of years.

Although any life on Titan might have a quite unfamiliar biochemistry, it would not be impossible to detect. To be effective, a search for such life would have to take a broader approach than screening for specific molecules, like DNA, ATP, or chlorophyll, which are central to terrestrial life. A life-searching probe on Titan would need to look for *any* complex organic compounds, especially those of high molecular weight (500 or greater). This is because we expect all life, whatever its origin, to depend on large, highly-specialised molecules. There are also good reasons to suppose that even life very different from our own would use biomolecular building blocks of a particular handedness. For example, all life on Earth uses right-handed sugars (in DNA, RNA, and metabolic pathways) and left-handed amino acids within proteins. Life elsewhere might differ in its preference for right-handed or left-handed versions of whatever molecules it depends on, but the chances are it will *have* a preference, which robotic instruments could be designed to detect.

Scientists are also aware of other likely signatures of life in general. For example, it's reasonable to assume, even in alien microbes, the presence of compartments and boundaries within a cell and from the cell to the outside environment to ensure that an organism can operate and maintain itself as an independent entity. A boundary layer is also needed in order to regulate the intake of food from the environment and prevent the uptake of harmful substances. Its nature may be quite different from cell membranes used by living things on Earth, but it would serve the same essential purpose.

The metabolism of Earth organisms is based on an intricate interplay of reduced and oxygenated compounds, in other words compounds that have either a surplus or a deficiency of electrons. Astrobiologists suspect this may be a universal condition of life. Since Titan's environment has a huge over-abundance of reducing chemicals, the discovery of oxygenated compounds in a high enough quantity might indicate the presence of chemical reactions mediated by life.

Until recently, scientists tended to shy away from talking about actual biology on Titan, preferring to entertain only the possibility of pre-biological chemistry. But new findings about the moon by the Cassini probe have encouraged bolder statements. For example, a report on the limits of organic life in planetary systems published in 2007 by the National Academy of Sciences (USA) pointed out that Titan has the basic requirements for life, including disequilibrium conditions, abundant carbon-containing compounds, and a fluid environment. "This makes inescapable," it said, "the conclusion that if life is an intrinsic property of chemical reactivity, life should exist on Titan."[72] Future missions to Saturn and its giant moon will put this intriguing assessment to the test.

## Target Titan

Cassini-Huygens, which arrived at Saturn in 2004, has sent back stunning images of Titan and vastly increased our knowledge of this extraordinary world. Now a new mission is needed to explore further Titan's mounting biological credentials. A proposal has been put together by more than 100 scientists and engineers, led by Athena Coustenis from the Observatoire de Meudon in France, called the Titan and Enceladus Mission (TANDEM) or Titan Saturn System Mission (TSSM). TANDEM would focus on Titan as a comparable environmental system to Earth, particularly to early Earth, and examine Titan's huge inventory of organic compounds to see whether some of these compounds are of a pre-biological or biological nature. Thus, this mission, planned for launch around 2020, would be aimed at finding some of the potential life indicators we talked about in the previous section. If these indicators are found, a follow-up mission to Titan would be sent to confirm the presence of life.

One of the advantages of TANDEM is that a similar strategy could be used in the case of Saturn's moon Enceladus. Sampling organic compounds from Enceladus is more easily accomplished than from Titan, because hot water eruptions catapult these compounds into space from which they can be gobbled up and analysed by an orbiter, making a lander unnecessary.

TANDEM would take seven to eight years to reach Saturn. On arrival, it would go into orbit around the gas giant and drop off a hot-air balloon just prior to its first Titan flyby. Buoyed by Titan's thick lower atmosphere, the balloon would drift around collecting data over a six-month period and relaying it back to Earth via the orbiter's telecommunications system. The lander, targeted at Kraken Mare, a northern lake consisting of petroleum-like compounds, would be released during the second Titan flyby and

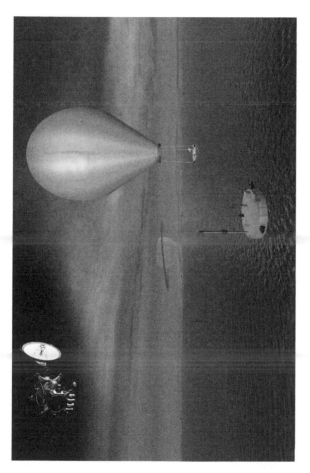

**Figure 28** Artist's depiction of a possible scenario for a future mission to Titan, as conceived for the Titan Saturn System Mission (TSSM), a joint NASA/ESA proposal for the exploration of Saturn and its moons. While a Titan-dedicated orbiter provides global remote science, context information, and relaying communications to and from Earth, a comprehensive *in situ* investigation is accomplished via a hot-air balloon circumnavigating Titan at altitudes between 2 km and 10 km and a probe with the capability to land on the surface of a northern lake to study the liquid composition

transmit images and measurements for at least nine hours (compared with the planned mere 20 minutes of the Huygens lander). During a two-year Saturn tour phase, the TANDEM orbiter would perform seven close flybys of Enceladus as well as 16 Titan flybys. Finally, the orbiter would descend into Titan's atmosphere and start sampling the gases and compounds present in the big moon's gassy envelope (see figure 28).

# 9

# Bioverse

Several hundred planets have been discovered around other stars since the first was detected back in 1988.[73] Many of these "exoplanets" are gas giants like Jupiter and Saturn and a good number of them circle their host stars at remarkably small distances (in some cases, much less than the distance of Mercury from the Sun). This is not because big planets in tiny orbits are disproportionately common but rather that these types of worlds are the easiest to detect, given that one of the most used search methods involves looking for slight wobbles in the motion of stars caused by planets moving around them. As detection methods become more sensitive and diverse, and are run over longer periods of time, smaller and smaller extrasolar worlds are coming to light. This trend is expected to continue until we know about large numbers of exoplanets similar in size to the Earth.

## Terrae Novae

Not only are solitary planets being found, but also planetary systems, containing two, three, or more worlds. An especially intriguing example is the system of Gliese 581, a star which, at a distance of just 20.4 light-years, is among the one hundred closest stars to Earth. Gliese 581 is an M-type dwarf, commonly known as a red dwarf. It is much cooler and fainter than the Sun, and only about one third as massive.

Gliese 581 has been a recent focus of attention by astronomers because, of the four known planets which move around it, three are probably rocky in nature and two of these orbit within the so-called habitable zone of the star. One way to define a habitable zone is the region around a star in which a planet must move in order to sustain liquid water on its surface. The planet known as Gliese 581c weighs in at just over five Earths, whereas Gliese 581d is slightly more hefty, at seven times the Earth's mass. Both are of astrobiological interest. Gliese 581e, which orbits much closer to the parent star, is, at the time of writing, the lightest exoplanet known with a mass only twice that of our own world.

Another lightweight extrasolar planet is MOA-2007-BLG-192Lb, which is about three times as heavy as Earth. The "MOA" part of its name is an acronym for Microlensing Observations in Astrophysics, an international project to detect remote objects, including exoplanets, using gravitational microlensing. In this technique, the presence of an unseen object is revealed by the way it bends and magnifies light from a background source, such as a far-away star or galaxy. Microlensing is ideally suited to searching for planets around stars that lie far outside the solar neighbourhood. The distance to MOA-2007-BLG-192Lb is an astounding 3,000 light-years.

Although MOA-2007-BLG-192Lb isn't much bigger than Earth, it exists in a very different stellar environment. The parent object around which it moves, at a distance similar to that of Venus from the Sun, is not an ordinary yellow star like the Sun, or even a much dimmer red dwarf, but a *brown* dwarf, with only about one-twentieth of the Sun's mass. A brown dwarf is effectively a failed star, too small to have properly ignited its core nuclear energy reserves. The planets of a brown dwarf receive little in the way of heat and light from the outside. However, under the right conditions, such worlds could have a sizable supply of internal heat, which might be enough, as in the case of Jupiter's moon Europa, to sustain a liquid ocean beneath the surface.

## A question of habitability

Worlds in unfamiliar situations, such as MOA-2007-BLG-192Lb, give us yet another reason to rethink the notion of where life might be found. When the concept of stellar habitable zones was first closely considered, in the 1960s, the emphasis was on what conditions were necessary for liquid water to exist on the surface of a planet. At the time it seemed a sensible approach given that, on Earth, water is intimately connected with life, and that surface water has been plentiful here for billions of years. But recently, from studies of large moons in the outer Solar System, such as Europa, Ganymede, and Titan, the strong possibility has emerged that vast amounts of liquid water can exist underneath the icy crust of some worlds. As a result, scientists have begun to look for less parochial ways to set limits on where life might be found.

To begin with, however, let's stick with the traditional approach to figuring out habitable zones in order to come up with some conservative estimates. If we do so, then the inner edge of the Sun's habitable zone – the minimum distance at which liquid water could endure on a planetary surface – is between about 0.84 and 0.95 astronomical units (AU), where 1 AU is the average distance of the Earth from the Sun. By this reckoning Venus lies outside the inner edge of the Sun's habitable zone today, although in the early solar system, when the Sun was less bright, Venus would have fallen inside the habitable zone. The outer edge of the habitable zone of the Solar System extends to at least 1.43 AU and, possibly, as much as 2 AU. Estimates vary because the limits of the habitable zone depend on a number of factors, including the makeup and density of a planet's atmosphere, the planetary mass, and how much energy the planet reflects back into space. Depending on the assumptions made, Mars, at an average distance of 1.52 AU from the Sun, may lie comfortably within the outer edge of the solar habitable zone, or just outside it.

If our solar system is typical, then many stars with planetary systems probably have at least one planet orbiting within their habitable zones. This should be true even if the central star is not a G-type star like the Sun. A G-type star appears white to yellow, is about the same size and mass as the Sun, and produces energy by converting hydrogen to helium in its core by nuclear fusion. If the star is less massive and less luminous (like a red dwarf) the habitable zone is closer to the star; if it is heavier and brighter (like an F-type star) the habitable zone is further away.

Just because an alien world lies within its star's habitable zone by no means guarantees, of course, that it will develop life. Also of great importance is an atmosphere dense enough to allow liquid water to be stable and also afford some protection from the perils of space. Such atmospheres, however, appear to be common, judging by planets (and even one moon) in our own Solar System and early measurements of atmospheres of extrasolar worlds. (Substantial atmospheres, of at least water vapour, would probably exist around Europa and Ganymede if these moons were closer to the Sun.)

Another crucial property, if a planet or moon is to be conducive to life, is the presence of some kind of dynamic activity either on or below the surface. Volcanic eruptions, earthquakes, and landslides often bring death and destruction to human communities, but such events, in a broader context, are characteristic of a living world. Our closest neighbour in space, the Moon, is inert both biologically and geologically. By contrast, the Earth is vibrantly alive in both these senses, its geological dynamism fuelled mainly by plate tectonics – the horizontal and vertical movement of solid plates over a more plastic interior. Carbon is buried, thus averting a runaway greenhouse effect of the type that has afflicted Venus, and nutrients and raw materials are continuously recycled. Dynamic activity on Titan and also to some degree on Mars add to the biological credentials of both these places.

An adequate mass for a planet or moon is also an asset for life, because a larger planet usually has a larger inventory of radioactive elements, which provide heat that can drive dynamic processes, such as plate tectonics. At some point the radioactive elements that heat the interior become depleted, and the smaller the planet the faster this depletion occurs. To date, the mass and radioactive endowment of Earth have been sufficient to keep plate tectonics going as a major recycling mechanism. But Mars has not been so lucky. Mars likely had plate tectonics early in its history – for the first 500 million years or so. But then the radioactive storehouse ran out and any volcanic activity on the Red Planet nowadays is localised, mostly in the region of Mount Olympus, the tallest volcano and mountain in the solar system. Had Mars been more massive, it might still have had enough internal heat to power plate tectonics and thus been more suited to life at present.

Earth is the largest terrestrial planet in the solar system, but there are certainly larger ones going around other stars. Gliese 581c and 581d seem to be terrestrial – meaning they are made predominantly of rock rather than gas – and the second largest terrestrial planet in our solar system (Venus) is virtually the same size as Earth. Terrestrial planets have a size limit, however, and an exoplanet that has at least ten times the Earth's mass is most likely to be a gas giant.

Gas giants have a distinctive type of atmosphere, as we know from the members of this planetary class in the solar system. We even have a rough idea of the atmospheric makeup of some extrasolar gas giants. Thanks to specially-equipped orbiting observatories, scheduled to be launched over the next decade or so, we shall soon have the means to detect gases such as water vapour, nitrogen, carbon dioxide, and oxygen around terrestrial exoplanets. Based on current knowledge there's every reason to believe that Earth-class planets are common around other stars and that analysis of their atmospheres will reveal many of them to be habitable.

## If and when life emerges

The big question though is whether, if a planet is habitable, life actually comes about on it. This is a much thornier issue. We know the environmental conditions that have supported life on Earth throughout the aeons, and those environmental conditions range widely enough to suggest that life forms on Earth would be able to live on many planetary bodies elsewhere in the universe. However, the appropriate conditions for the origin of life are much less well known and it might be that these conditions are a tiny subset of those that would allow the continued presence of life.

How can we tell if a new-found world is inhabited or not? If life is thoroughly interwoven with a planet and as abundant as on Earth, certain molecules, such as chlorophyll and DNA, might give away its presence. Certain atmospheric compositions, too, such as the simultaneous presence of methane and oxygen, would provide tantalising hints of biogenicity. The situation would be less clear, however, in the case of planets where life is scarce and ekes out an existence in small, isolated refuges or in the subsurface. Hunting down such residual life on an otherwise barren world could prove to be a daunting challenge. The organisms might dwell in underground regions to which our remote sensing technology is blind. Martian life, if it exists, might well be an example of this hard-to-find subsurface biology. The only life-detection strategy likely to succeed in such circumstances would be to visit the world in question, do experiments on or below ground level, and, if possible, bring samples back to Earth for lab analysis. But even negative results wouldn't prove that a world was lifeless, because creatures might be hiding in niches that we hadn't thought to sample or didn't have the means to reach.

What should be our approach where only *in situ* investigations are likely to be effective? The issue of contamination becomes paramount. One could argue that any indigenous life would be much better adapted to the local environmental conditions than any organisms arriving onboard a spacecraft and would therefore prevail. While this is probably true in general, experience from Earth with animal and plant species crossing continents provides plenty of counterexamples. In any event, the decision to land equipment on other, possibly life-bearing worlds, is as much an ethical one as a technical one. If there are other intelligent beings in the galaxy that face or have faced this decision, some may have opted only to observe remotely worlds on which life might have taken hold, while others may have been happy to send in their robot explorers. Some races might go even further and intentionally seed planets and moons that are capable of supporting non-native life. There is even the extraordinary possibility that we ourselves are the product of such a seeding program, as the late Nobel laureate Francis Crick and the late Leslie Orgel once contemplated.[74]

We know surprisingly little about the origin of life on our own planet. Even the environmental conditions under which life on Earth first emerged are unknown. Suggestions for the birthplace of terrestrial life have ranged from warm ponds or lagoons to ice and hydrothermal vents, but each idea has its shortcomings. Until we know how life evolved on Earth and the circumstances under which it might do so on other planets, we'll lack a firm theoretical basis on which to predict how common life might be across the universe. One of the few useful clues we have is that life on Earth is ancient and has left traces in some of the earliest preserved terrestrial rocks, some four billion years old. This suggests that life can get off the ground quite easily, so that it may be a pretty common phenomenon.

## Life around red dwarfs?

M-type dwarfs, or red dwarfs, are (with the exception of brown dwarfs) by far the commonest stellar denizens of the universe. As many as three-quarters of all stars fall into this category. And although they are dim by solar standards, their longevity is prodigious. Whereas the Sun has a total lifespan as a normal or "main sequence" star of about eight billion years before it heats up and expands dramatically to become a red giant (rendering the Earth uninhabitable), red dwarfs eke out their core hydrogen fuel reserves for more than 50 billion years. This compares with a current age of the universe of a mere 13 billion years. Such extreme longevity has far-reaching implications for intelligent life in particular, because any intelligence that had emerged under the feeble, cool rays of a red dwarf star when the universe was young could by now be vastly more advanced than human beings.

Not only are red dwarfs abundant and long-lived, but their luminosity, or rate of energy output, remains more-or-less steady over many billions of years. The result is a habitable zone that changes very little over the lifetime of the star. This is in marked contrast to the Sun. In the early days of the solar system, the Sun was about 30%

dimmer than it is now. Gradually its energy output increased, and this trend will continue into the future until, some four billion years from now, the Sun becomes unstable and turns into a red giant. The solar radius will expand beyond Earth's orbit, destroying not only all terrestrial life but the Earth itself. In fact our planet will become uninhabitable long before this final catastrophe. About one billion years from now the Sun will get so hot that it will trigger a runaway greenhouse effect on Earth which will cause the oceans to boil away. Earth will endure the same fate that befell Venus at least a billion years ago. The comparative long-term stability of a red dwarf clearly offers great advantages to any life that might have evolved nearby.

On the other hand, the dimness of a red dwarf means that its habitable zone is quite small – typically between 10% and 40% of the Earth-Sun distance. This could be detrimental from a biological perspective because any planet orbiting within the habitable zone experiences such an enormous gravitational pull from its sun that it easily becomes tidally locked, meaning that the time it takes to complete one orbit (its year) is the same as the time it needs to spin once around it axis (its day). Tidal locking is responsible for the Moon always keeping the same face toward the Earth. Any planet that had one hemisphere constantly facing toward the central star, and the other always facing away, would experience permanent temperature extremes that might make it difficult for life to develop.

However, since there are so many red dwarfs, there must be many planets orbiting within their habitable zones which are not tidally locked. In fact, without any special attempt to go out looking for them, a good number of red dwarf planets (Gliese 581c and d among them) have already been found. Theoretical modelling also indicates that super-Earths, rocky like the Earth but heavier, can easily form during the formation of planetary systems.[75] Massive Earth-like planets would be much less likely to succumb to tidal locking.

Another potential hurdle to life in the vicinity of red dwarfs is their ultraviolet and X-ray emissions. These can be particularly strong in young red dwarfs. High-energy radiation is generally harmful to life, but there are two important caveats. First, some scientists believe that ultraviolet light may have played a key role in chemical reactions important to the origin of life on Earth. One intriguing example is the formation of the amino acid glycine from three molecules of hydrocyanic acid in water when exposed to ultraviolet radiation. This reaction also produces cyanamide, a compound that joins amino acids together to form proteins. Furthermore, adding two more molecules of hydrocyanic acid and adenine can form an important compound in the biochemistry of cellular respiration, protein and ATP synthesis, and the structure of DNA and RNA.

While ultraviolet radiation is a challenge to life (for example, humans develop sunburn, and even skin cancer, if we are exposed to too much of it), it clearly has also been an important factor in biological evolution. If life on Earth hadn't been constantly challenged it would not be so superbly adapted to its environment and would probably have remained at a bacterial stage. Ultraviolet radiation triggers mutations, which are mostly harmful to the individual organism. But a few mutations (usually less than one per cent) are beneficial in that they propel life on to new evolutionary trajectories.

Most planetary atmospheres provide natural protection against high-energy radiation. Even the thin atmosphere on Mars prevents hard (high-frequency) ultraviolet or X-rays from reaching ground level (although enough ultraviolet does get through to pose a real threat to any organisms exposed on the surface). Terrestrial life today is largely shielded from ultraviolet radiation by ozone in the higher atmosphere and from cosmic radiation (harder radiation) by its magnetic field, though there was clearly life on Earth before an oxygen-rich atmosphere or ozone layer developed. Before the ozone shield formed, life was restricted to water, because water is a good absorber of ultraviolet radiation. Only when the ozone barrier was in place could the land areas of our world be conquered by creatures that crawled out of the sea. Then biological ultraviolet blockers, such as fur and special pigments (including melanin, found in human skin) evolved.

Based on current evidence, it seems that planets around red dwarfs are common and a good number of these planets should be capable of harbouring life. In fact, judging by how long – at least four billion years – it took for complex life (animals and plants) and, in particular, intelligent life to develop on our own world, it appears that red dwarfs have certain advantages. The relatively brief lifespan of G stars like the Sun, and the shifting habitable zone during this time, makes the evolution of higher life-forms on any accompanying planets a race against time. The fairly constant environment offered by red dwarfs, interspersed with occasional ultraviolet flares to push life in new directions, would seem better suited to the nurturing of complex life. Added to the prospects for advanced indigenous life is the possibility of intelligent life that evolved around G stars migrating to the planets of red dwarfs in search of long-term security.

## Kepler

As you read this, a remarkable spacecraft, launched by NASA in March 2009, is trailing around the Sun in the same orbit as Earth, with a telescope pointed on a small patch of sky in the constellation of Cygnus. Its goal is to look systematically at about 90,000 stars for signs of Earth-like planets moving around them. Over a four-year period it will give us a good idea of how common other Earths are in our galaxy, and, by implication, how common life-as-we-know-it might be.

The Kepler spacecraft, named after the great German astronomer who lived from 1571 to 1630, will focus specifically on planets within the habitable zones of their central stars. By the end of its mission, we should know of hundreds and perhaps thousands of new worlds on which other instruments can be trained to gather more information. Both orbiting and ground-based observatories will be able to target these new-found planets to image them, analyse their atmospheric compositions, and, ultimately, determine if it seems likely that they harbour life. Even as we seek confirmation that our own solar system is inhabited, we are looking much further afield – out to distances of several hundred light-years – for traces of biology around other suns.

# Conclusion

**W**e believe that the Viking landers found life on Mars and that more recent evidence, especially from Martian meteorites and the discovery of methane on the Red Planet, strongly supports this claim. Proof that is acceptable to the scientific community at large will come only from further robotic and possibly manned missions which can carry out more sophisticated biological tests *in situ* or return samples of Martian soil to Earth for detailed analysis. This proof will probably be available sometime between 2020 and 2030.

We would also be very surprised if life had not evolved on our other near planetary neighbour, Venus, in the remote past. We think there's a reasonable chance that this life adapted to drastically changing conditions on the second planet and is extant today in the lower cloud deck, giving rise to the intriguing spectra and particle types of this region.

In the case of both Mars and Venus, there is the distinct possibility that some cross-fertilisation took place with our own world. Meteorites may well have carried viable microbes between the inner planets of the solar system between about 3.5 and 4 billion years ago. If so, astrobiologists will be faced with the tricky task of determining on which of the rocky worlds life originated and to which it was later transplanted by rocks splashed into space during asteroid collisions.

Such a problem is not likely to arise in the case of planets and moons lying beyond the main asteroid belt. We believe that one or more of the Galilean satellites, most notably Europa, along with Titan and Enceladus of Saturn, are outstanding candidates for life of a very different kind to that found on Earth. Alien microbes, and possibly larger organisms, likely exist, we suspect, in liquid bodies of water or hydrocarbons on these faraway worlds. If they do, there is virtually no chance that they are in any way related to life in the inner solar system. They would be independent instances of biology, possibly based on entirely different molecular building blocks.

Astrobiologists would no longer be surprised if life in general turns out to be common throughout the universe. We've given, in this book, good reasons to suppose that we are not alone even in our own little neck of the cosmic woods. The discovery of plentiful planets around other stars promises a multitude of places for biology to take hold throughout the Milky Way Galaxy and beyond. But obviously the commonest organisms, cosmos-wide, are going to be primitive, just as they are on Earth. Simple things always outnumber more complex ones. A question of immense importance, not only to scientists but to humanity as a whole, is how often basic life forms – on a level, say, with bacteria – evolve to become elaborate, multicellular, macroscopic, and, eventually, intelligent beings.

We are not alone, we Earth-dwellers, even within the Sun's domain. A challenge for the future will be to discover if there are fellow races and civilisations in the Galaxy with whom we may learn to communicate.

# References

1. Ezell, E. C. and Ezell, L. N. (1984) *On Mars: Exploration of the Red Planet 1958–1978*. NASA History Office NASA SP-4212, Scientific and Technical Information Branch, National Aeronautics and Space Administration, Washington, D.C.

2. Horowitz, N. H., Cameron, R.E., and Hubbard, J. S. (1972) Microbiology of the Dry Valleys of Antarctica. *Science* 176: 242–245.

3. Levin, G. V., and Straat, P. A. (1977) Recent results from the Viking labeled release experiment on Mars. *Journal of Geophysical Research* 82: 4663–4667.

4. Klein, H. P. (1978) The Viking biological experiments on Mars. *Icarus* 34: 666–674.

5. Benner, S. A., Devine, K. G., Matveeva, L. N., and Powell, D. H. (2000) The missing organic molecules on Mars. *Proceedings of the National Academy of Science* USA 97: 2425–2430.

6. Navarro-González, R., Navarro, K. F., de la Rosa, J., Iniguez, E., Molina, P., Miranda, L. D., Morales, P., Cienfuegos, E., Coll, P., Raulin, F., Amils, R., and McKay, C.P. (2006) The limitations on organic detection in Mars-like soil by thermal volatilization – gas chromatography-MS and their implications for the Viking results. *Proceedings of the National Academy of Science* USA 103: 16089–16094.

# REFERENCES

7. Moore, H. J., Hutton, R. E., Scott, R. F., Spitzer, C. R., and Shorthill, R. W.(1977) Surface materials of the Viking landing sites. *Journal of Geophysical Research* 82: 4497–4523.
8. Schulze-Makuch, D., Irwin, L. N., Lipps, J. H., LeMone, D., Dohm, J. M., and Fairén, A. G. (2005) Scenarios for the evolution of life on Mars. Special Edition on Early Mars of *Journal of Geophysical Research – Planets* 110: E12S23, doi:10.1029/2005JE002430.
9. Houtkooper, J. M. and Schulze-Makuch, D. (2007) A possible biogenic origin for hydrogen peroxide on Mars: the Viking results reinterpreted. *Int. J. of Astrobiology* 6: 147–152.
10. Levin, G. V., and Straat, P. A. (1977) Recent results from the Viking labeled release experiment on Mars. *Journal of Geophysical Research* 82: 4663–4667; Levin, G. V., Straat, P. A. (1981) A search for a nonbiological explanation of the Viking Labeled Release Life Detection Experiment. *Icarus* 45: 494–516.
11. McKay, D. S., Gibson, E. K., Thomas-Keprta, K. L., Vali, H., Romanek, C. S., Clemett, S. J., Chillier, X. D. F., Maechling, C. R., and Zare, R. N. (1996) Search for past life on Mars: possible relic biogenic activity in Martian meteorite ALH84001. *Science* 273: 924–930.
12. Miura, Y. N., Nagao, K., Sugiura, N., Sagawa, H., and Matsubara, L. (1995) Orthopyroxenite ALH84001 and shergottite ALH77005: Additional evidence for a Martian origin from noble gases. *Geochimica et Cosmochimica Acta* 59: 2105–2113.
13. Folk, R. L. (1993) SEM imaging of bacteria and nanobacteria in carbonate sediments and rocks. *Journal of Sedimentary Research* 63: 990–999; Folk, R. L. (1999) Nannobacteria and the precipitation of carbonate in unusual environments. *Sedimentary Geology* 126: 47–55.
14. Ciftçioglu, N. Björklund, M., Kuorikoski, K., Bergström, K., and Kajander, E. O. (1999) Nanobacteria: an infectious cause for kidney stone formation. *Kidney International* 56: 1893–1898; Kajander, E. O. and Ciftçioglu, N. (1998) Nanobacteria: an alternative mechanism for pathogenic intra- and extracellular calcification and stone formation. *Proceedings of the National Academy of Science* USA 95: 8274–8279; Kajander, E. O., Kuronen, I., Akerman, K., Pelttari, A., and

176

Ciftçioglu, N. (1997) Nanobacteria from blood, the smallest culturable autonomously replicating agent on Earth. *Proceedings of SPIE* 3111: 420–428.

15. Baker, B.J., Tyson, G.W., Webb, R.I., Flanagan, J., Hugenholtz, P., and others (2006) Lineages of acidophilic archaea revealed by community genomic analysis. *Science* 314: 1933–1935.

16. Eiler, J. M., Valley, J. W., Graham, C. M., and Fournelle, J. (2002) Two populations of carbonate in ALH84001: geochemical evidence for discrimination and genesis. *Geochim. et Cosmochim. Acta* 66: 1285–1303; Romanek, C. S., Grady, M. M., Wright, I. P., Mittlefehldt, D. W., Socki, R. A., Pillinger, C. T., and Gibson, E. K. (2002) Record of fluid-rock interactions on Mars from meteorite ALH84001. *Nature* 372: 655–657.

17. Thomas-Keprta, K. L., Bazylinski, D. A., Kirschvink, J. L., Clemett, S. J., McKay, D. S., Wentworth, S. J., Vali, H., Gibson, E. K., and Romanek, C. S. (2000) Elongated prismatic magnetite crystals in ALH84001 carbonate globules: potential Martian magnetofossils. *Geochim. et Cosmochim. Acta* 64: 4049–4081.

18. Thomas-Keprta, K. L., Clemett, S. J., Bazylinski, D. A., Kirschvink, J. L., McKay, D. S., Wentworth, S. J., Vali, H., Gibson, E. K., McKay, M. F., and Romanek, C. S. (2001) Truncated hexa-octahedral magnetite crystals in ALH84001: presumptive biosignatures. *Proceedings of the National Academy of Science* USA 98: 2164–2169.

19. Friedmann, E. I. Wierzchos, J., Ascaso, C., and Winklhofer, M. (2001) Chains of magnetite crystals in the meteorite ALH84001: evidence of biological origin. *Proceedings of the National Academy of Science* USA 98: 2176–2181.

20. Bradley, J. P., Harvey, R. P., and McSween, H. Y. (1996) Magnetite whiskers and platelets in the ALH84001 Martian meteorite: evidence of vapor phase growth. *Geochim. Cosmochim. Acta* 60: 5149–5155.

21. Barber, D. J. and Scott, E. R. D. (2002) Origin of supposedly biogenic magnetite in the Martian meteorite Alan Hills 84001. *Proceedings of the National Academy of Science* USA 99: 6556–6561.

22. Thomas-Keprta, K. L., Clemett, S. J., Bazylinski, D. A., Kirschvink,

J. L., McKay, D. S., Wentworth, S. J., Vali, H., Gibson, E. K., and Romanek, C.S. (2002) Magnetofossils from ancient Mars: a robust biosignature in the Martian meteorite ALH84001. *Applied and Environmental Microbiology*, August 2002: 3663–3672.

23. Weiss, B. P., Sam Kim, S., Kirschvink, J. L., Kopp, R. E., Sankaran, M., Kobayashi, A., and Komeili, A. (2004) Magnetic tests for magnetosome chains in Martian meteorite ALH84001. *Proceedings of the National Academy of Sciences* USA 101: 8281–8284.

24. McKay, D. S. and 10 co-authors (2006) Analysis of *in situ* carbonaceous matter in Martian meteorite Nakhla. *Astrobiology* 6: 184; McKay, D. S., Clemett, S. J., Thoomas-Keprta, K. L., Wentworth, S. J., Gibson, E. K., Robert, F., Verchovsky, A. B., Pillinger, C. T., Rice, T., and Van Leer, B. (2006) Observation and analysis of *in situ* carbonaceous matter in Nakhla: part I. 37th Lunar and Planetary Science Conference, Houston, Texas, abstract # 2251.

25. Fisk, M. R., Popa, R., Mason, O. U., Storrie-Lombardi, M. C., and Vicenzi, E.P. (2006) Iron-magnesium silicate bioweathering on Earth (and Mars?). *Astrobiology* 6: 48–68.

26. Furnes, H., Banerjee, N. R., Muehlenbachs, K., Staudigel, H., and de Wit, M. (2004) Early life recorded in Archean pillow lavas. *Science* 304: 578–581.

27. Cockell, C. S., Schuerger, A. C., Billi, D., Friedmann, E. I., and Panitz, C. (2005) Effects of a simulated Martian UV flux on the cyanobacterium, *Chroococcidiopsis* sp. 029. *Astrobiology* 5: 127–140.

28. Diaz, B. and Schulze-Makuch, D. (2006) Microbial survival rates of *E. coli* and *D. radiodurans* under single and combined stresses of temperature, pressure, and UV radiation, and its relevance to Martian environmental conditions. *Astrobiology* 6: 332–347.

29. Friedmann, E. I. (1982) Endolithic microorganisms in the Antarctic cold desert. *Science* 215: 1045–1053.

30. Fairén, A. G., Dohm, J. M., Öner, T., Ruiz, J., Rodríguez, A. P., Schulze-Makuch, D., Ormö, J., McKay, C. P., Baker V. R., and Amils, R. (2004) Updating the Evidence of Oceans on Early Mars. Early Mars 2004 Conference, Jackson, Wyoming, 11–15 October

2004; Ruiz, J., Fairén, A. G., Dohm, J. M., and Tejero, R. (2004) Thermal isostasy and deformation of possible paleoshorelines on Mars. *Planetary and Space Science* 52: 1297–1301.

31. Perron, J. T., Mitrovica, J. X., Manga, M., Matsuyama, I., and Richards, M. A. (2007) Evidence for an ancient Martian ocean in the topography of deformed shorelines. *Nature* 447: 840–843.

32. Dohm, J. M., Baker, V. R., Boynton, W. V., Fairén, A. G., Ferris, J. C., Finch, M., Furfaro, R., Hare, T. M., Janes, D. M., Kargel, J. S., Karunatillake, S., Keller, J., Kerry, K., Kim, K., Komatsu, G., Mahaney, W. C., Schulze-Makuch, D., Marinangeli, L., Ori, G. G., and Ruiz, J. (2009) GRS evidence and the possibility of paleooceans on Mars. *Planetary Space and Science* 57: 664–684

33. Montmessin, F. (2006) The orbital forcing of climate changes on Mars. *Space Science Reviews* 125: 457–472.

34. Schulze-Makuch, D., Irwin, L. N., Lipps, J. H., LeMone, D., Dohm, J. M., and Fairén, A. G (2005) Scenarios for the evolution of life on Mars. Special Edition on Early Mars of *Journal of Geophysical Research – Planets* 110: E12S23, doi:10.1029/2005JE002430.

35. Cano R. J., Borucki, M. (1995) Revival and identification of bacterial spores in 25 to 40 million year old Dominican amber. *Science* 268: 1060–1064.

36. Vreeland, R. H., Rosenzweig W.D., and Powers D. W. (2000) Isolation of a 250 million-year-old halotolerant bacterium from a primary salt crystal. *Nature* 407: 897–900.

37. Malin, M. C. and Edgett, K .S. (2000) Evidence for recent ground-water seepage and surface runoff on Mars. *Science* 288: 2330–2335.

38. Malin, M. C, Edgett, K. S., Posiolova,L. V., McColley, S. M., and Noe Dobrea, E. Z. (2006) Present-day impact cratering rate and contemporary gully activity on Mars. *Science* 314: 1573–1577.

39. Krasnopolski, V. A., Maillard, J. P., and Owen, T. C. (2004) Detection of methane in the Martian atmosphere: evidence for life? *Icarus* 172, 537–547; Formisano, V., Atreya, S., Encrenaz, T., Ignatiev, N. and Giuranna, M. (2004). Detection of methane in the atmosphere of Mars. *Science* 306, 1758 – 1761; Mumma, M. J.,

Novak, R. E., DiSanti, M. A., Bonev, B. P., and Dello Russo, N. (2004) Detection and mapping of methane and water on Mars. American Astronomical Society, DPS meeting #36, *Bulletin of the American Astronomical Society* 36, 1127.

40. Nair, J., Summers, M. E., Miller, C. E., and Yung, Y. L. (2005) Isotopic fractionation of methane in the Martian atmosphere. *Icarus* 175: 32–35.

41. Oehler, D. Z., Allen, C. C. McKay, D. S. (2005) Impact metamorphism of subsurface organic matter on Mars: a potential source for methane and surface alteration. Presentation at 36th Lunar and Planetary Science Conference, Houston, Texas, 14–18 March 2005.

42. Onstott, T. C., McGown, D., Kessler, J., Lollar, B. S., Lehmann, K. K., and Clifford, S. M. (2006) Martian $CH_4$: Sources, Flux, and Detection. *Astrobiology* 6: 377–295.

43. Stevens, T. O. and McKinley, J. P. (1995) Lithoautotrophic microbial ecosystems in deep basalt aquifers. *Science* 270: 450–455.

44. See the News Release by the journal *Nature* in October of 2008: http://www.nature.com/news/2008/081022/full/4551018a.html).

45. Schulze-Makuch, D., Fairén, A.G., and Davila, A. F. (2008) The case for life on Mars. *Int. J. of Astrobiology* 7: 117–141.

46. Hecht, M.H. and 13 co-authors (2009) Detection of perchlorate and the soluble chemistry of Martian soil at the Phoenix lander site. *Science* 325: 64–67.

47. Boynton, W.V. and 13 co-authors (2009) Evidence for calcium carbonate at the Mars Phoenix landing site. *Science* 325: 61–64.

48. Smith, P.H. and 35 co-authors. (2009) $H_2O$ at the Phoenix landing site. *Science* 325: 58–61; Renno, N. and 22 co-authors (2009) Physical and thermodynamical evidence for liquid water on Mars. Lunar and Planetary Science Conference XL (abstract # 1440).

49. Ehlmann, B. L. and 14 other authors (2008) Orbital identification of carbonate-bearing rocks on Mars. *Science* 322: 1828–1832.

50. Whiteway, J.A. and 23 co-authors (2009) Mars water-ice clouds and precipitation. *Science* 325: 68–70.

51. Dohm, J. M., Baker, V. R., Boynton, W. V., Fairén, A. G., Ferris,

J. C., Finch, M., Furfaro, R., Hare, T. M., Janes, D. M., Kargel, J. S., Karunatillake, S., Keller, J., Kerry, K., Kim, K., Komatsu, G., Mahaney, W. C., Schulze-Makuch, D., Marinangeli, L., Ori, G. G., and Ruiz, J. (2009) GRS evidence and the possibility of paleooceans on Mars. *Planetary Space and Science* 57: 664–684.

52. See the News Release by the journal *Nature* in October of 2008: http://www.nature.com/news/2008/081022/full/4551018a.html).

53. Allen, C.C. and Oehler, D.Z. (2008) A case for ancient springs in Arabia Terra, Mars. *Astrobiology* 8: 1093–1112.

54. Houtkooper, J. M. and Schulze-Makuch, D. (2007) The hydrogen peroxide-water hypothesis for life on Mars and the problem of detection. Proceedings of SPIE – Volume 6694, *Instruments, Methods, and Missions for Astrobiology X*, R. B. Hoover, G. V. Levin, A.Y. Rozanov, P. C. W. Davies (eds), 66940N; Schulze-Makuch, D., Fairén, A. G., and Davila, A. F. (2008) The case for life on Mars. *Int. J. of Astrobiology* 7: 117–141.

55. Schulze-Makuch, D., Turse, C., Houtkooper, J. M., and McKay, C. P. (2008) Testing the $H_2O_2$-$H_2O$ hypothesis for life on Mars with the TEGA instrument on the Phoenix lander. *Astrobiology* 8: 205–214.

56. Schulze-Makuch, D. and Irwin, L. N. (2004 hard cover, 2006 soft cover) *Life in the Universe: Expectations and Constraints*. Springer, Berlin, 172 p.; Schulze-Makuch, D. and Irwin, L.N. (2008) 2nd edition, *Life in the Universe: Expectations and Constraints*. Springer, Berlin, 251 p.

57. Schleper, C., Peuhler, G., Holz, I., Gambacorta, A., Janekovic, D., Santarius, U., Klenk, H. P., and Zillig, W. (1996) *Picrophilus* gen. Nov., fam. Nov.: a novel aerobic, heterotrophic, thermoacidophilic genus and family comprising Archaea capable of growth around pH 0. *J. Bacteriol.* 177: 7050–7079.

58. Sattler, B., Puxbaum, H., and Psenner, R. (2001) Bacterial growth in supercooled cloud droplets. *Geophys. Res. Lett.* 28: 239–242.

59. Schulze-Makuch, D. and Irwin, L. N. (2002) Reassessing the possibility of life on Venus: Proposal for an Astrobiology Mission. *Astrobiology* 2: 197–202; Cockell, C.S. (1999) Life on Venus. *Planet.*

*Space Science* 47: 1487–1501; Grinspoon, D.H. (1997) *Venus Revealed: A New Look Below the Clouds of our Mysterious Twin Planet*, Perseus Publishing, Cambridge, MA.

60. Schulze-Makuch, D., Grinspoon, D. H., Abbas, O., Irwin, L. N. and Bullock, M. (2004) A sulfur-based UV adaptation strategy for putative phototrophic life in the Venusian atmosphere. *Astrobiology* 4: 11–18.

61. Schulze-Makuch, D., Irwin, L. N. and Irwin, T. (2002) Astrobiological relevance and feasibility of a sample collection mission to the atmosphere of Venus. ESA Special Publication SP-518, p. 247–252.

62. Lewis, J. S. (1971) Satellites of the outer planets: their physical and chemical nature. *Icarus* 15, 174–185.

63. Muller, A.W. J. (1985) Thermosynthesis by biomembranes: energy gain from cyclic temperature changes. *Journal of Theoretical Biology* 115: 429–453; Muller, A. W. J. (2003) Finding extraterrestrial organisms living on thermosynthesis. *Astrobiology* 3: 555–564; Muller, A. W. J. and Schulze-Makuch, D. (2006) Thermal energy and the origin of life. *Origin of Life and Evolution of Biospheres* 36: 177–189.

64. Schulze-Makuch, D. and Irwin, L. N. (2002) Energy cycling and hypothetical organisms in Europa's ocean. *Astrobiology* 2: 105–121; Schulze-Makuch, D. and Irwin, L. N. (2008) 2nd *Life in the Universe: Expectations and Constraints* (2nd edition) Springer, Berlin, 251 p.

65. Irwin, L. N. and Schulze-Makuch, D. (2003) Strategy for modeling putative ecosystems on Europa. *Astrobiology* 3 (Special Issue on Europa): 813–821.

66. Daniel, R.M., Dunn, R.V., Finney, J.L., and Smith, J.C. (2003) The role of dynamics in enzyme activity. *Ann. Rev. Biophys. Biomol. Struct.* 32: 69–92; Bragger, J.M.. Dunn, R.V., and Daniel, R.M. (2000) Enzyme activity down to −100°. *Biochimica Biophysica Acta* 1480: 278–282.

67. Lunine, J. I., Yung, Y. L., and Lorenz, R. D. (1999) On the volatile inventory of Titan from isotopic substances in nitrogen and methane. *Planetary Space and Science* 47: 1291–1303.

68. Schulze-Makuch, D. and Grinspoon, D. H. (2005) Biologically enhanced energy and carbon cycling on Titan? *Astrobiology* 5: 560–567.

69. McKay, C. P. and Smith, H. D. (2005) Possibilities for methanogenic life in liquid methane on the surface of Titan. *Icarus* 178: 274–276.

70. Abbas, O. and Schulze-Makuch, D. (2002) Acetylene-based pathways for prebiotic evolution on Titan. *ESA Special Publication* SP-518: 345–348; Schulze-Makuch, D. and Irwin, L. N. (2008) 2nd *Life in the Universe: Expectations and Constraints* (2nd edition) Springer, Berlin, 251 p.

71. Bains, W. (2004) Many chemistries could be used to build living systems. *Astrobiology* 4: 137–167.

72. Baross, J. A., Benner, S. A., Cody, G. D., Copley, S. D., Pace, N. R. et al (2007) *The Limits of Organic Life in Planetary Systems*. National Academies Press, Washington, D.C.

73. Campbell, B., Walker, G.A.H., and Yang, S. (1988) A search for substellar companions to solar-type stars. *Astrophysical Journal* 331: 902–921.

74. Crick, F. H. C. and Orgel, L. E. (1973) Directed Panspermia. *Icarus* 19: 341–346.

75. Boss, A. P. (2006) Rapid formation of super-Earths around M dwarf stars. *Astrophysical Journal* 644: L79-L82.

# Index

# INDEX